PROTEIN FOLDING DISORDERS
From basic biology to public policy

Rodrick Wallace
Deborah Wallace
Division of Epidemiology
The New York State Psychiatric Institute
rodrick.wallace@gmail.com

November 15, 2011

Preface

Protein folding disorders of aging like Alzheimer's (AD) and Parkinson's diseases present intractable medical challenges: drug treatments have, at best, palliative impact, failing to alter ultimate disease course. The design of individual or population level interventions thus requires a deeper understanding of protein folding and its regulation than provided by contemporary 'physics' or culture-bound medical magic bullet models. This is particularly true in light of the relentless 'inverse Moore's Law' of drug development that characterizes the six-fold increase in the inflation-adjusted cost of bringing a drug to market since 1950, reaching \$ 1.2 billion by 2010. Consequently, the pharmaceutical industry has greatly reduced research on a range of poorly-understood afflictions, including AD. That is, in all likelihood, there will not be effective drugs for many protein folding disorders anytime soon, and if developed, the costs will surely be prohibitive. Here we will follow protein folding and its dysfunction from the cellular to social levels of organization, finding a strong foundation for effective public health policies against early disease onset.

Chapters 1 and 2 apply formal topological rate distortion methods to protein folding and regulation in a manner similar to Tlusty's (2010a) elegant exploration of the genetic code.

Chapter 3 generalizes the approach, finding large-scale, quasi equilibrium 'resilience' states representing normal and pathological protein folding regulation under a cellular-level cognitive paradigm similar to that proposed by Atlan and Cohen (1998) for the immune system.

Chapter 4 extends the argument, producing diffusion models of protein folding disorders in which epigenetic or life history factors determine the rate of onset of regulatory failure. Basically, this is a premature aging driven by familiar synergisms between disjunctions of resource allocation and need in the context of socially or physiologically toxic exposures and chronic powerlessness at individual and group scales. The work is consonant with recent results on psychosocial stress and premature aging (Epel et al., 2004).

In that spirit, Chapter 5 applies a culturally/environmentally-tuned HPA axis model to observed differences in Alzheimer's onset rates in

White and African American subpopulations as a function of an index of distress-proneness. More detailed modeling suggests two broad classifications of protein folding disorders, those representing the overloading of regulatory systems, and a parallel with certain forms of autoimmune disorders, in which a pathological state is misperceived as 'normal'. Further study suggests, given the underlying cognitive nature of protein folding regulation, that psychosocial stress can impair virtually any therapeutic intervention.

Chapter 6 examines recent trajectories of massive, policy-driven, deurbanization and deindustrialization in the US from this perspective, and provides two compelling models relating policy to Alzheimer's mortality incidence in the young elderly at the state level. As Chapter 7 indicates, these policies constitute a powerful 'anti-public health' case history for protein folding and other stress-induced disorders that, paradoxically, provides something of a roadmap for positive public health interventions.

The material is written at an advanced undergraduate mathematical level, requiring some familiarity with probability theory. New ideas are introduced as needed, and tutorials provided. The two public health chapters are heavily data-based, and will benefit from some degree of scientific sophistication.

The book should be of interest to a wide spectrum of mathematically-inclined students, researchers, and other stake-holders interested in protein folding, folding disorders, public policy, and public health.

On first reading, Sections 1.1, 5.5, and Chapters 6 and 7 form a concise nonmathematical introduction that can be extended as needed by the formal developments.

The material is a synthesis across several recent papers that use Tlusty's approach to examine protein folding:

Wallace, R., 2010, A rate distortion approach to protein symmetry, *BioSystems*, 101:97-108.

Wallace, R., 2010, A scientific open season: Comment, *Physics of Life Reviews*, 7:377-378.

Wallace, R., 2010, Structure and dynamics of the 'protein folding code' inferred using Tlusty's topological rate distortion approach, *BioSystems*, 103:18-26..

Wallace, R., 2010, Protein folding disorders: Toward a basic biological paradigm, *Journal of Theoretical Biology*, 267:582-594.

The cover illustration has been adapted from:

Kumar, S. and J. Udgaonkar, 2010, Mechanisms of amyloid formation by proteins, *Current Science*, 98:639-656.

Contents

Chapter 1

Protein symmetries

1.1   Introduction

At this writing, front page articles in major news and scientific media trumpet the intractability of protein folding disorders with headlines like "No Magic Bullet Against Alzheimer's Disease" (Kolata, 2010; O,Connor et al., 2010). Medical magic bullets are, of course, a Western, and indeed particularly American, cultural conceit. Heine (2001), describing a similar paradigm within psychology, writes

> The extreme nature of American individualism suggests that a psychology based on late 20th century American research not only stands the risk of developing models that are particular to that culture, but of developing an understanding of the self that is peculiar in the context of the world's cultures...

Henrich et al. (2010) have elaborated this point in an instantly famous critique titled "The Weirdest people in the world?". Given the fundamental biological nature of protein folding itself, a magic bullet perspective on it's disorders may be analogously weird, and inference based on American perceptions of current research similarly suspect (e.g., Kolata, 2010). Qui et al. (2009), based in Stockholm, present a markedly different view:

> Alzheimer's dementia is a multifactorial disease in which older age is the strongest risk factor... [that] may partially reflect the cumulative effects of different risk and protective factors over the lifespan, including the complex interactions of genetic susceptibility, psychosocial factors, biological factors, and environmental exposures experienced over the lifespan.

Qiu et al. (2009) explain that mutation effects account for only a small fraction of observed cases, and that the APOE ϵ4 allele – the only established genetic factor for both early and late onset disease – is a *susceptibility* gene, neither necessary nor sufficient for disease onset. They further describe how many of the same factors implicated in diabetes and cardiovascular disease predict onset of Alzheimer's as well: tobacco use, high blood pressure, high serum cholesterol, chronic inflammation, as indexed by a higher level of serum C-reactive protein, and diabetes itself. Effective protective factors include high educational and socioeconomic status, regular physical exercise, mentally demanding activities, and significant social engagement.

Qui et al. conclude:

> Epidemiological research has provided sufficient evidence that vascular risk factors in middle-aged and older adults play a significant role in the development and progression of dementia and [Alzheimer's disease], whereas extensive social network and active engagement in mental, social, and physical activities may postpone the onset of the dementing disorder. Multidomain community intervention trials are warranted to determine to what extent preventive strategies toward optimal control of multiple vascular factors and disorders, as well as the maintenance of an active lifestyle, are effective against dementia and [Alzheimer's disease].

Similarly, Fillit et al. (2008) find that lifestyle risk factors for cardiovascular disease, such as obesity, lack of exercise, smoking, and certain psychosocial factors, have been associated with an increased risk for cognitive decline and dementia, concluding that current evidence indicates an association between hypertension, dyslipidemia and diabetes and cognitive decline and dementia.

Here we will take what Heine and colleagues might describe as an 'East Asian' approach to protein folding disorders, and examine their embedding context from a new perspective, as opposed to perceiving them as separated from their backcloth, and hence 'naturally' amenable to magic bullets.

High rates of protein folding and aggregation diseases, in conjunction with observations of the elaborate cellular folding regulatory apparatus associated with the endoplasmic reticulum and other cellular structures that compare produced to expected protein forms (e.g., Scheuner and Kaufman, 2008; Dobson, 2003), presents a clear and powerful logical challenge to simple physical 'folding funnel' free energy models of protein folding, as compelling as these are *in vitro* or *in silico*. This suggests that a more biologically-based model is needed for understanding the life course trajectory of protein folding, a model analogous to Atlan and Cohen's (1998) cognitive paradigm for the immune system. That is, the intractable set of disorders related to protein aggregation and misfolding belies simple mechanistic approaches, although free energy landscape pictures (Anfinsen, 1973; Dill et al., 2007) surely capture part of the process (but see Chou and Carlacci, 1991). The diseases range from prion illnesses like Creutzfeld-Jakob disease, in addition to amyloid-related dysfunctions like Alzheimer's, Huntington's and Parkinson's diseases, and type 2 diabetes. Misfolding disorders include emphysema and cystic fibrosis.

The role of epigenetic and environmental factors in type 2 diabetes has long been known (e.g., Zhang et al., 2009; Wallach and Rey, 2009). Haataja et al. (2008), for example, conclude that the islet in type 2 diabetes shows much in common with neuropathology in neurodegenerative diseases where interest is now focused on protein misfolding and aggregation and the diseases are now often referred to as unfolded protein diseases.

Scheuner and Kaufman (2008) likewise examine the unfolded protein response in β cell failure and diabetes. Indeed, their opening paragraph raises the fundamental questions regarding the adequacy of simple energy landscape models of protein folding:

> In eukaryotic cells, protein synthesis and secretion are precisely coupled with the capacity of the endoplasmic reticulum (ER) to fold, process, and traffic proteins to the cell surface. These processes are coupled through several signal transduction pathways collectively known as the unfolded protein response [that] functions to reduce the amount of nascent protein that enters the ER lumen, to increase the ER capacity to fold protein through transcriptional upregulation of ER chaperones and folding catalysts, and to induce degradation of misfolded and aggregated protein.

Goldschmidt et al. (2010) describe pathological protein fibrillation as follows:

> We found that [protein segments with high fibrillation propensity] tend to be buried or twisted into unfavorable conformations for forming beta sheets... For some proteins a delicate balance between protein folding and misfolding exists that can be tipped by changes in environment, destabilizing mutations, or even protein concentration...
>
> In addition to the self-chaperoning effects described above, proteins are also protected from fibrillation during the process of folding by molecular chaperones...
>
> Our genome-wide analysis revealed that self-complementary segments are found in almost all proteins, yet not all proteins are amyloids. The implication is that chaperoning

effects have evolved to constrain self-complementary segments from interaction with each other.

Many of these processes and mechanisms seem no less examples
of chemical cognition than the immune/inflammatory responses that
Atlan and Cohen (1998) describe in terms of an explicit cognitive
paradigm, or that characterizes well-studied neural processes.

We will use Tlusty's (2007a, b, 2008a, b, c, 2010a) analysis of the
emergence of the genetic code as a basis for an appropriate model, and
begin, from the ground up, with a reconsideration of protein symmetry
from his perspective.

1.2   Protein symmetries

There are, it seems, numerous underlying 'protein folding codes' in the
sense of characteristic segments of amino acids whose ultimate folded
structures are a somewhat debatable matter of formal taxonomy. Figure 1.1, from Hartl and Hayer-Hartl (2009), schematically expands the
spectrum of final protein conformations according to an *in vivo* 'folding
funnel' model dispersed across a measure of intra- vs. inter-molecular
contact for hydrophobic-core proteins forming tertiary structure. Intra-
molecular conformations involve three-dimensional assemblages of α-
helices and β-sheets, while the most densely packed inter-molecular
form is, perhaps, the ubiquitous semicrystalline amyloid fibril.

The basic spectrum of figure 1.1 for proteins having a hydrophobic core, in general, explains the necessity of the elaborate regulatory
structures associated with the endoplasmic reticulum and its attendant
spectrum of chaperone proteins (e.g., Scheuner and Kaufman, 2008),
and the evolutionary pattern of protein sequences inferred by Goldschmidt et al. (2010). The inevitable corrosion of the cellular regulatory apparatus with age would then explain the subsequent onset of
amyloid fibril and other aggregation disorders.

Most particularly, the spectrum of valleys in figure 1.1 characterizes
a set of equivalence classes that defines a 'protein folding groupoid', in
the sense of Weinstein (1996). As we will argue below, both the native
state and amyloid fibril have structured subdivisions, internal equivalence classes, that define a nested set of groupoids. See the Mathematical Appendix for a summary of standard material on groupoids.

With regard to the disjunction between 'native' and 'amyloid' protein forms, very early on, Astbury (1935) conjectured that globular
proteins could also have a linear state, based on pioneering x-ray studies. Chiti et al. (1999) argue that

> ...[P]rovided appropriate conditions are maintained over
> prolonged periods of time, the formation of ordered amyloid

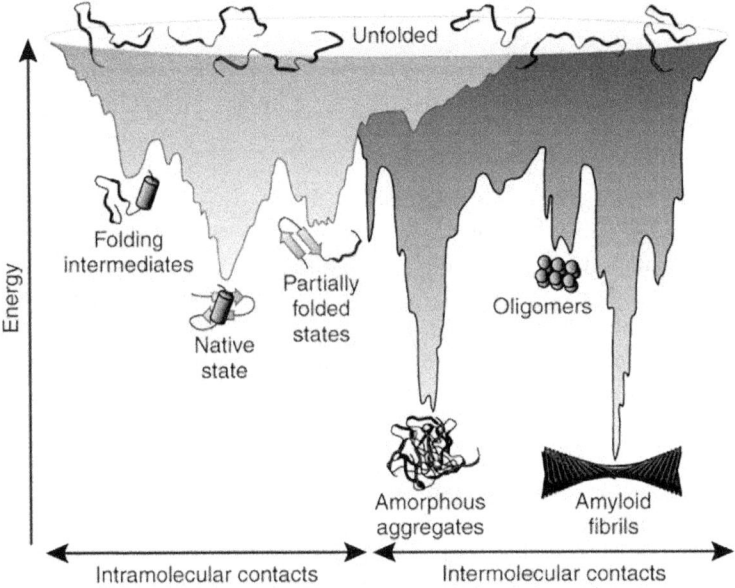

Figure 1.1: From Hartl and Hayer-Hartl, (2009). Energy landscape spectrum of protein folding and aggregation, parsed according to the degree of intra- vs. inter-molecular contact. Each energy valley defines an equivalence class, and the set of such classes defines the 'protein folding groupoid', in the sense of Weinstein (1996). Four basic classifications can be seen; native state, amorphous aggregates, semi-structured oligomers, and quasi-crystalline amyloid fibrils. Within the native state and the amyloid fibrils, systematic subclasses can be identified, leading to a fine structure for protein coding.

protofilaments and fibrils could be an intrinsic property of many polypeptide chains, rather than being a phenomenon limited to a very few aberrant sequences.

Wang et al. (2008), in a an elegant series of experiments on bacterial inclusion bodies, conclude that

> ..[A]myloid aggregation appears to be a common property of protein segments and consequently is observed in both eukaryotes and prokaryotes... [Thus] there must be evolved strategies against amyloid formation, which include both quality control mechanisms through molecular chaperones as well as sequence-based [evolutionary] prevention of amyloid aggregation...
>
> ...[E]ach protein may exist, not only in an unfolded or folded state, but, by containing at least one amino acid segment that is capable of participating in a sequence-specific, ordered, cross-β-sheet aggregated state, may also exist in an amyloid-like aggregate. The process of protein aggregation can thus be viewed as a primitive folding mechanism, resulting in a defined, aggregated conformation with each aggregated protein having its own distinctive properties.

Krebs et al. (2009), however, in a paper tellingly titled 'Protein aggregation: more than just fibrils', find that the amyloid fibril is not the only structure that aggregating proteins of widely different types may adopt. For example, the occurrence of spherulites, which have been found *in vivo* as well as *in vitro*, appears to be generic, although the factors that determine the equilibrium between free fibril and spherulite are not as yet clear. That is, we have not fully explained the spectrum implied by figure 1.1. Nevertheless, here we will use Tlusty's (2007a, b, 2008a, b, c, 2010a) arguments on the evolution of the genetic code to explore something of that spectrum.

As Kamtekar et al. (1993) point out, experimental studies of natural proteins show how their structures are remarkably tolerant to amino acid substitution, but that tolerance is limited by a need to maintain the hydrophobicity of interior side chains. Thus, while the information needed to encode a particular protein fold is highly degenerate, this degeneracy is constrained by a requirement to control the locations of polar and nonpolar residues. This is the precise protein folding analog to Tlusty's error network analysis of the genetic code, and his graph coloring arguments should thus apply, in some measure, to protein folding as well, allowing inference on the underlying structure of the 'protein folding codes' to be associated with the horizontal axis of figure 1.1.

Tyco (2006), likewise, argues that the amyloid fibril is a generically stable structural state of a polypeptide chain, competing thermodynamically and kinetically with globular monomeric states and unfolded monomeric states. Peptides and proteins that are known to form amyloid fibrils have widely diverse amino acid sequences and molecular weights. He particularly finds that

> The near sequence independence of amyloid formation represents a challenge to our understanding of the physical chemistry of peptides and proteins.

Such sequence independence is, again, very precisely the degeneracy associated with Tlusty's error network approach.

Intermediate forms in figure 1.1 remain to be studied from this perspective.

Some of these matters have, of course, already been the subject of considerable attention. A series of elegant experiments by the Hecht group (e.g., Hecht et al., 2004), extending the work of Kamtekar et al. (1993), has focused on a basic understanding of protein folding through substitution of different polar and nonpolar amino acids in the construction of normal and fibril proteins. α-helices are found to be natural outcomes of amino acid sequences having a 3.6 residue/turn patten, i.e., a digital signal of the form 101100100110, where 1 indicates a polar, and 0 a nonpolar amino acid. The resulting three dimensional structures are formed by the propensity of the different residues to interact with an aqueous environment.

β sheets, on the other hand, emerge from a simpler period 2 code, e.g., 1010101, matching the structural repeat of the sheets. More recent work (Kim and Hecht, 2006) finds that generic hydrophobic residues of this form are sufficient to promote aggregation of the Alzheimer's Aβ42 peptide. However, while the positioning of hydrophobic residues is more important than the exact identities of the hydrophobic side chains for determining overall geometry, reaction kinetics, the rate of fibril formation, was profoundly affected by those identities. This suggests that the 'protein folding code' may be, in no small part, contextual, that is, determined as much by *in vivo* cellular regulatory machinery as by *in vitro* hydrophobic/hydrophilic physical interactions. This, we will suggest below, likely involves the operation of something like the catalytic mechanisms that Wallace and Wallace (2009) and Wallace (2010a) describe.

1.2.1 Large scale structure

Broadly, figure 1.1 embraces a four-fold classification (Wallace, 2010b):

1. The 'native state' determined, at low concentrations, entirely by the amino acid sequence in the classic sense of Anfinsen (1973).

2. Amorphous aggregates.

3. Semi-structured oligomers, as explored by Krebs et al. (2009).

4. Amyloid/amyloid-like one-dimensional fibrils.

Following the description by Tlusty, (2010b), the genetic code is a mapping of one codon to one amino acid. By contrast, the 'protein folding code' is a mapping of genes to folded amino acid chains, and the complexity gap between the two codes is very great indeed (e.g., Mirny et al., 2001). The strategy that allows adaptation of Tlusty's methods to protein folding is a coarse-graining of protein structure into a matrix of larger building blocks, e.g., α-helices and β-sheets. At this lower resolution a 'code' is a mapping between short DNA stretches, analogous to codons, and the convoluted motifs of proteins, playing the role of amino acids. As a consequence of the great tolerance to amino acid substitutions described above, as long as charge and polarity are conserved, it is possible to cluster all the sequences that encode the same structural motif. This greatly reduces the size of the resulting DNA sequence graph and thus limits the number of possible building blocks.

Tlusty's method (Tlusty, 2010c) is analagous, but not identical, to the well-known topological coloring problem. In the coding problem, one desires maximum similarity in the colors of neighboring 'countries', while in the coloring problem one must color neighboring countries by different colors.

Generalizing Table 1 of Tlusty (2007, 2010a) according to the genus γ of the underlying graph, that is, the number of holes in the error network associated with the proposed code, we can apply Heawood's graph genus formula for the coloring number that identifies the maximal number of first excited modes of the coding graph Laplacian,

$$chr(\gamma) = Int[1/2(7 + \sqrt{1 + 48\gamma})].$$

(1.1)

where Int is the integer value of the enclosed expression and γ itself is defined from Euler's formula (Tlusty, 2010) as

$$\gamma = 1 - \frac{1}{2}(V - E + F)$$

(1.2)

where V is the number of code network vertices, E the number of network edges, and F the number of enclosed faces.

Equation (1.1) produces the table

γ (# network holes)	chr(γ) (# prot. syms.)
0	4
1	7
2	8
3	9
4	10
5	11
6, 7	12
8, 9	13

In Tlusty's scheme, the second column represents the maximal possible number of product classes that can be reliably produced by error-prone codes having γ holes in the underlying coding error network.

From Tlusty's perspective, then, our four-fold classification for figure 1.1 produces a the simplest possible large-scale 'protein folding code', a sphere limited by the four-color problem, and the simplest cognitive cellular regulatory system would thus be constrained to pass/fail on four basic flavors, as it were, of folded proteins.

Within the funnel leading to the native state, however, chaperone processes would face far more difficult choices.

This suggests a possible two-fold cellular regulatory structure, and next we consider the two most fully characterized geometric structures in more detail, the normal and amyloid forms.

1.2.2 Normal globular proteins

Normal irregular protein symmetries were first classified by Levitt and Chothia (1976), following a visual study of polypeptide chain topologies in a limited dataset of globular proteins. Four major classes emerged; all α-helices; all β-sheets; α/β; and $\alpha + \beta$, as illustrated in figure 1.2.

While this scheme strongly dominates observed irregular protein forms, Chou and Maggiora (1998), using a much larger data set, recognize three more 'minor' symmetry equivalence classes; μ (multidomain); σ (small protein); and ρ (peptide), and a possible three more 'subminor' groupings.

We infer that the normal globular 'protein folding code error network' is, essentially, a large connected 'sphere' – producing the four

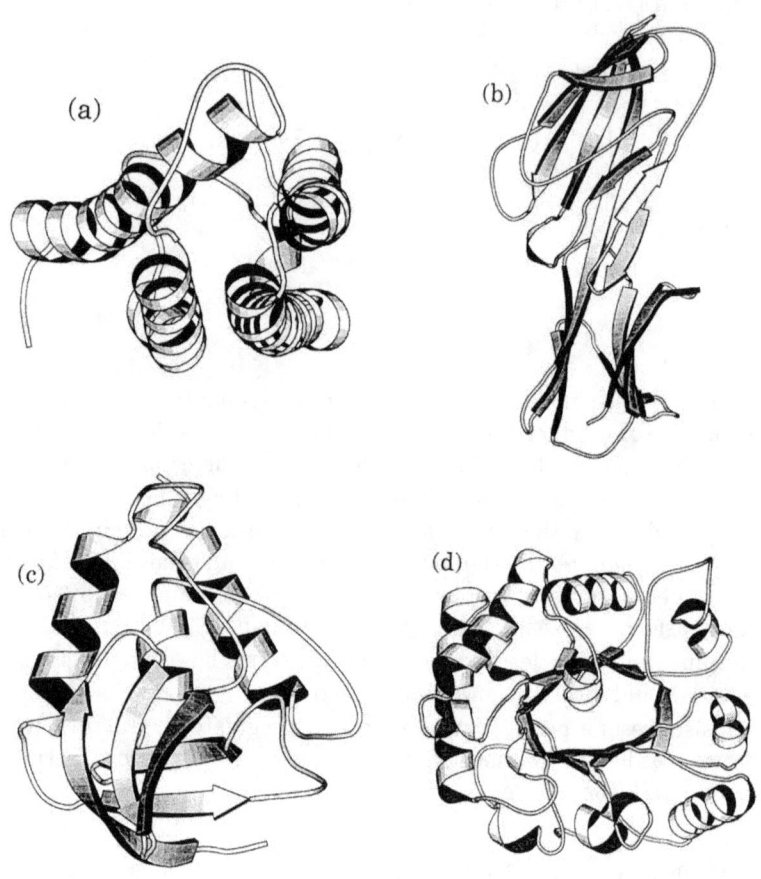

Figure 1.2: From Chou and Zhang, (1995). Standard equivalence classes for inexact protein symmetries according to Levitt and Chothia, 1976: (a) All-α helices. (b) All-β sheets. (c) $\alpha + \beta$. (d) α/β. More recent work identifies a minimum of seven, and possibly as many as ten, such classes (Chou and Maggiora, 1998).

dominant structural modes of figure 1.2 – having one minor, and possibly as many as three more 'subminor' attachment handles, in the Morse Theory sense (Matsumoto, 2002), a matter opening up other analytic approaches.

1.2.3 Amyloid fibrils

As described above, Kim and Hecht (2006) suggest that overall amyloid fibril geometry is very much driven by the underlying β-sheet coding 1010101, although the rate of fibril formation may be determined by exact chemical constitution. Work by Sawaya et al. (2007) parses some of those subtleties: They identify an eight-fold 'steric zipper' symmetry necessarily associated with the linear amyloid fibrils that characterize a vast spectrum of protein folding disorders. Figure 1.3, adapted from their work, shows those symmetries. In essence, two identical sheets can be classified by the orientation of their faces (face-to-face/face-to-back), the orientation of their strands (with both sheets having the same edge of the strand up or one up and the other down), and whether the strands within the sheets are parallel or anti parallel. Five of the eight symmetry possibilities have been observed. This suggests, from the text table above, that the 'amyloid folding code error network' is a double donut, that is, has two, different sized, interior holes, resembling, perhaps, a toroid with a smaller attachment handle.

1.2.4 Amyloid self-replication

Maury (2009) has recently proposed an 'amyloid world' model for the emergence of prebiotic informational entities, based on the extraordinary stability of amyloid structures in the face of the harsh conditions of the prebiotic world. From this perspective, the synthesis of RNA, and the evolution of the RNA-protein world, were later, but necessary events for further bimolecular evolution. Maury further argues that, in the contemporary DNA$\Leftrightarrow$RNA$\Rightarrow$protein world, the primordial β-conformation-based information system is preserved in the form of a cytoplasmic epigenetic memory.

Falsig et al. (2008) examine the many different strains of prions, finding that differences in kinetics of the elementary steps of prion growth underlie the differential proliferation of prion strains, based on differential frangibility of prion fibrils. They argue that an important factor is the size of the stabilizing cross-β amyloid core that appears to define the physical properties of the resulting structures, including their propensity to fragment, with small core sizes leading to enhanced frangibility. In terms of the protein folding funnel approach, they find that intrinsic frustration implies that several distinct arrangements favoring a certain subset of globally incompatible interactions are possible,

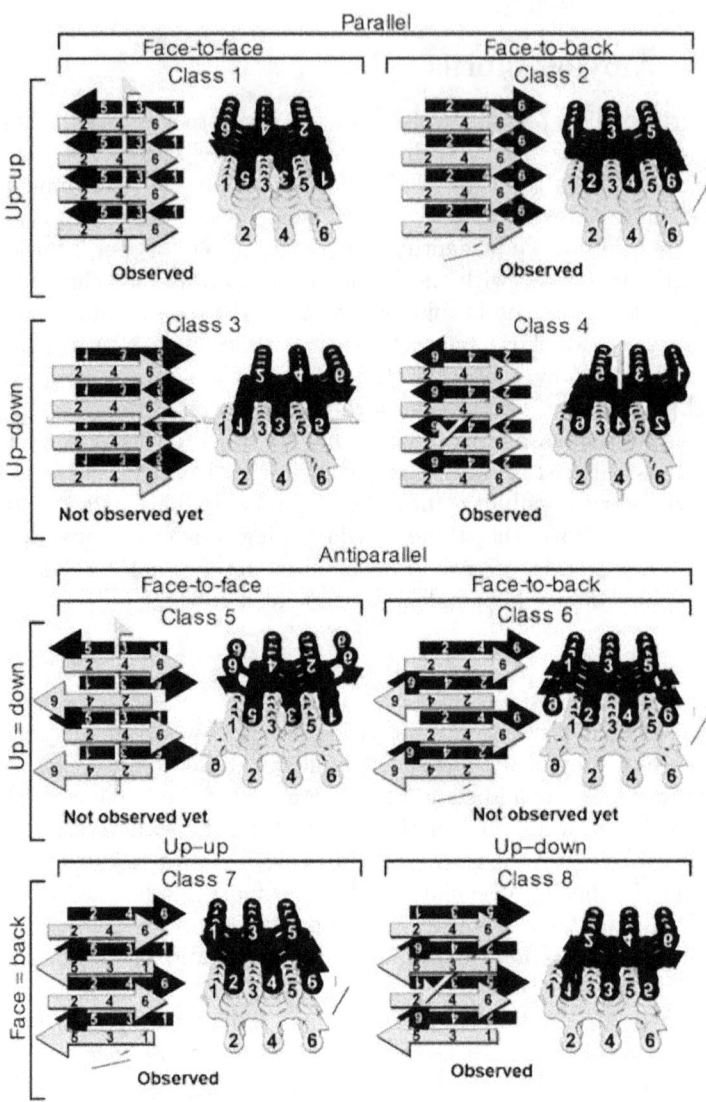

Figure 1.3: From Sawaya et al., (2007). The eight possible steric zipper symmetry classifications for amyloid fibrils.

reflecting the observed strain-dependent differences in the parts of the sequence incorporated into the fibril core.

In addition, they argue, there are unexplored similarities between Alzheimer's and prion diseases, that is, the analogies between prion and Aβ aggregates could be broader than initially suspected.

Given the eight-fold symmetry of the amyloid fiber, say versions A → H, then the simplest 'frangibility code' is the set of identical pairings:
$$\{AA, BB, ..., GG, HH\},$$
producing eight different possible structures and their reproduction by fragmentation. More complex prion symmetries, or the possibility of combinatorial recombination, would allow a much richer structure, producing quasi-species, in the sense of Collinge and Clarke (2007). Permitting different sequence lengths or explicitly identifying different sequence orders would vastly enlarge what Collinge has characterized as a 'cloud' of possibilities, in the case of prion diseases. Indeed, classic studies by Bruce and Dickinson (1987) found 15 or more different prion strains in a mouse model.

Recent work on prions appears to support something of Maury's hypothesis. Li et al. (2010) find that infectious prions, mainly what is called PrP^{Sc}, a spectrum of β sheet-rich conformers of the normal host protein PrP^C, undergo Darwinian evolution in cell culture. In that work, prions show the evolutionary hallmarks: they are subject to mutation, as evidenced by heritable changes of their phenotypes, and to selective amplification, as found by the emergence of distinct populations in different environments. Figure 1.4, from Li et al. (2010), shows a prion energy landscape similar to figure 1.1. This suggests the possibility of characterizing the underlying topology of a 'prion reproduction code', in the sense of the sections above.

One might speculate that prions and prion diseases represent fossilized remains of Maury's prebiotic amyloid world.

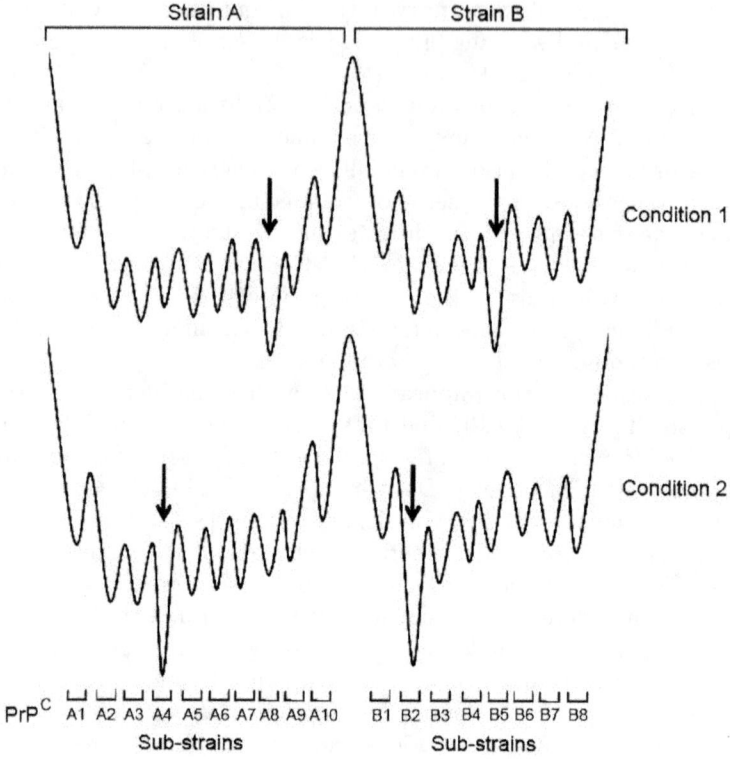

Figure 1.4: From Li et al. (2010), figure S10. Schematic energy land-scape for prion strains and substrains. The energy landscape diagram suggests that substrains are distinguishable collectives of prions that interconvert reproducibly and readily because they are separated by low activation energy barriers. The properties of a strain may vary depending on the environment in which it replicates, as the proportions of component substrains may change to favor that replicating most rapidly, indicated by the arrows. Comparison with figure 1.1, and the subsequent argument, suggests an underlying topological structure for a 'prion reproduction code'.

Chapter 2

Protein folding dynamics

2.1 Spontaneous symmetry breaking

We begin the theoretical analysis of protein folding dynamics with a classic conceptual context:

Landau's theory of phase transitions (Landau and Lifshitz, 2007) assumes that the free energy of a system near criticality can be expanded in a power series of some 'order parameter' ϕ representing a fundamental measurable quantity, that is, a symmetry invariant. One writes

$$F_0 = \sum_{k=m}^{p(>m)} A_k \phi^k,$$

with $A_2 \approx \alpha(T - T_c)$ sufficiently close to the critical temperature T_c. This mean field approach can be used to describe a variety of second-order effects for $p = 4$ or $p = 6$, $A_3 = 0$ and $A_4 > 0$, and first order phase transitions (requiring latent heat) for either $p = 6, A_3 = 0, A_4 < 0$ or $p = 4$ and $A_3 \neq 0$. These can be both temperature induced (for $m = 2$) and field induced (for $m = 1$).

Minimization of F_0 with respect to the order parameter yields the average value of ϕ, $< \phi >$, which is zero above the critical temperature and non-zero below it. In the absence of external fields, the second-order transition occurs at $T = T_c$, while the first-order, needing latent heat, occurs at $T_c^* = T_c + A_4^2/4\alpha A_6$. In the latter case thermal hysteresis arises between $T_s \equiv T_c + A_4^2/3\alpha A_6$ and T_c. A more accurate approximation involves an expression that recognizes the effect of coarse-graining, adding a term in $\nabla^2 \phi$ and integrating over space rather than summing. Regimes dominated by this gradient will show behaviors analogous to those described using the one dimensional Landau-Ginzburg equation, which, among other things, characterizes superconductivity.

The Landau formalism quickly enters deep topological waters (Pettini, 2007, pp. 42-43; Landau and Lifshitz, 2007, pp. 459-466). The essence of Landau's insight was that phase transitions without latent heat – second order transitions – were usually in the context of a significant symmetry change in the physical states of a system, with one phase, at higher temperature, being far more symmetric than the other. A symmetry is lost in the transition, a phenomenon called spontaneous symmetry breaking. The greatest possible set of symmetries in a physical system is that of the Hamiltonian describing its energy states. Usually states accessible at lower temperatures will lack symmetries available at higher temperatures, so that the lower temperature phase is the less symmetric: The randomization of higher temperatures ensures that higher symmetry/energy states will then be accessible to the system.

At the lower temperature an order parameter must be introduced to describe the system's physical states – some extensive quantity like magnetization. The order parameter will vanish at higher temperatures, involving more symmetric states, and will be different from zero in the less symmetric lower temperature phase.

This can be formalized as follows (Pettini, 2007). Consider a thermodynamic system having a free energy F which is a function of temperature T, pressure P, and some other extensive macroscopic parameters m_i, so that $F = F(P, T, m_i)$. The m_i all vanish in the most symmetric phase, so that, as a function of the m_i, $F(P, T, m_i)$ is invariant with respect to the transformations of the symmetry group G_0 of the most symmetric phase of the system when all $m_i \equiv 0$.

The state of the system can be represented by a vector $|m> = |m_1, ..., m_n>$ in a vector space $\mathcal{E}$. Now, within $\mathcal{E}$, construct a linear representation of the group G_0 that associates with any $g \in G_0$ a matrix $M(g)$ having rank n. In general, the representation $M(g)$ is reducible, and we can decompose $\mathcal{E}$ into invariant irreducible subspaces $\mathcal{E}_1, \mathcal{E}_2, ..., \mathcal{E}_k$, having basis vectors $|e_i^{(n)}>$ with $n = 1, 2, ...n_i$ and $n_i = dim\mathcal{E}_i$. The state variables m_i are transformed into new variables $\eta_i^{(n)} = <e_i^{(n)}|m>$, where the bracket represents an inner product.

In terms of irreducible representations $D_i(g)$ induced by $M(g)$ in $\mathcal{E}_i$ we have

$$M(g) = D_1(g) \oplus D_2(g) \oplus, ..., \oplus D_k(g).$$

If at least one of the $\eta_i^{(n)}$ is nonzero, then the system no longer has the symmetry G_0. This symmetry has been broken, and the new symmetry group is G_i, associated with the representation $D_i(g)$ in $\mathcal{E}_i$. The variables $\eta_i^{(n)}$ are the new order parameters, and the free energy is now $F = F(P, T, \eta_i^{(n)})$. For a physical system the actual values of the η as functions of P and T can be variationally determined by minimizing the free energy F.

Two essential features distinguish information systems, like the translation of a genome into a folded protein, from this simple physical model.

First, the dynamics of order parameters cannot always be determined by simplistic minimization procedures in biological circumstances (e.g., Levinthal, 1969): embedding environments can, within contextual constraints (that particularly include available metabolic free energy), write images of themselves via evolutionary selection mechanisms, driving the system toward such structures as the protein folding funnel (e.g., Levinthal, 1968; Wolynes, 1996).

Second, the essential symmetry of information sources is quite often driven by groupoid, rather than group, structures (e.g., Wallace and Fullilove, 2008). One must then engage the full transitive orbit/isotropy group decomposition, and examine groupoid representations (e.g., Bos, 2007; Buneci, 2003) configured about the irreducible representations of the isotropy groups. This observation seems particularly relevant given the usual helix/sheet/connecting loop tilings that characterize most elaborate protein conformations (Chou and Zhang, 1995).

2.2 Information theory

Here we think of the machinery listing a sequence of codons as communicating with machinery that produces amino acids, folds them in a particular real-world physiological context, and produces the final symmetric protein. We then suppose it possible to compare what is actually produced with what should have been produced – what the codon stream proposes, taking Anfinsen's (1973) perspective – and what the protein production machinery disposes – comparing observed folded proteins with their idealized image, that can now be well described using 'physics' models like Rosetta. Such comparison is entirely an empirical matter, and can be done by human experimenters, but is most often done in real time by internal cellular machinery: the endoplasmic reticulum and friends.

The average distortion between what is sent by the codon stream and what is observed in the cell or tissue is an essential parameter of the transmission channel, and the relation between the minimum channel capacity needed for some average distortion measure is a fundamental empirical characteristic of an information channel, characterized by the Rate Distortion Theorem, one of the basic asymptotic limit theorems of probability theory. The rate distortion function, $R(D)$, that measures the minimum channel capacity ensuring an average distortion D, by some measure is, in essence, a different way of looking at the protein folding funnel.

Onuchic and Wolynes (2004) have put something of the matter fully in evolutionary terms:

> Protein folding should be complex... a folding mechanism must involve a complex network of elementary interactions. However, simple empirical patterns of protein folding kinetics... have been shown to exist.
>
> This simplicity is owed to the global organization of the landscape of the energies of protein conformations into a funnel...This organization is not characteristic of all polymers with any sequence of amino acids, but is a result of evolution...
>
> Evolution achieves robustness by selecting for sequences in which the interactions present in the functionally useful structure are not in conflict, as in a random heteropolymer, but instead are mutually supportive and cooperatively lead to a low energy structure. The interactions are 'minimally frustrated'... or 'consistent'...

It is possible to reframe this mechanism in formal information theory terms.

Suppose a sequence of signals is generated by a biological information source Y having output $y^n = y_1, y_2, ...$ – codons. This is 'digitized' in terms of the observed behavior of the system with which it communicates, say a sequence of 'observed behaviors' $b^n = b_1, b_2, ...$ – amino acids and their folded protein structure. Assume each b^n is then deterministically retranslated – that is, 'decoded' in engineering jargon – back into a reproduction of the original biological signal, $b^n \to \hat{y}^n = \hat{y}_1, \hat{y}_2, ...$ To reiterate, such decoding can be done by human experimenters, taking Anfinsen's viewpoint that the codon stream characterizes the intended protein form as a message that is distorted by transmission along the DNA $\to$ RNA $\to$ Protein channel.

Define a distortion measure $d(y, \hat{y})$ that compares the original to the retranslated/decoded path. Many distortion measures are possible. The Hamming distortion is defined simply as

$$d(y, \hat{y}) = 1, y \neq \hat{y}$$

$$d(y, \hat{y}) = 0, y = \hat{y}.$$

For continuous variates the squared error distortion is just $d(y, \hat{y}) = (y - \hat{y})^2$.

There are many such possibilities. The distortion between *paths* y^n and $\hat{y}^n$ is defined as $d(y^n, \hat{y}^n) \equiv \frac{1}{n} \sum_{j=1}^{n} d(y_j, \hat{y}_j)$.

A remarkable fact of the Rate Distortion Theorem is that *the basic result is independent of the exact distortion measure chosen* (Cover and Thomas, 1991; Dembo and Zeitouni, 1998).

Suppose that with each path y^n and b^n-path retranslation into the y-language, denoted $\hat{y}^n$, there are associated individual, joint, and conditional probability distributions $p(y^n), p(\hat{y}^n), p(y^n, \hat{y}^n), p(y^n|\hat{y}^n)$.

The average distortion is defined as

$$D \equiv \sum_{y^n} p(y^n) d(y^n, \hat{y}^n).$$

(2.1)

It is possible, using the distributions given above, to define the information transmitted from the Y to the $\hat{Y}$ process using the Shannon source uncertainty of the strings:

$$I(Y, \hat{Y}) \equiv H(Y) - H(Y|\hat{Y}) = H(Y) + H(\hat{Y}) - H(Y, \hat{Y}),$$

where $H(...,...)$ is the standard joint, and $H(...|...)$ the conditional, Shannon uncertainties (Cover and Thomas, 1991; Ash, 1990).

If there is no uncertainty in Y given the retranslation $\hat{Y}$, then no information is lost, and the systems are in perfect synchrony.

In general, of course, this will not be true.

The *rate distortion function* $R(D)$ for a source Y with a distortion measure $d(y, \hat{y})$ is defined as

$$R(D) = \min_{p(y,\hat{y}); \sum_{(y,\hat{y})} p(y)p(y|\hat{y})d(y,\hat{y}) \leq D} I(Y, \hat{Y}).$$

(2.2)

The minimization is over all conditional distributions $p(y|\hat{y})$ for which the joint distribution $p(y, \hat{y}) = p(y)p(y|\hat{y})$ satisfies the average distortion constraint (i.e., average distortion $\leq D$).

The *Rate Distortion Theorem* states that $R(D)$ is the minimum necessary rate of information transmission that ensures the communication between the biological vesicles does not exceed average distortion D. Thus $R(D)$ defines a minimum necessary channel capacity. Cover and

Thomas (1991) or Dembo and Zeitouni (1998) provide details. The rate distortion function has been calculated for a number of systems.

We reiterate an absolutely central fact characterizing the rate distortion function: Cover and Thomas (1991, Lemma 13.4.1) show that $R(D)$ *is necessarily a decreasing convex function of* D *for any reasonable definition of distortion.*

That is, $R(D)$ *is always* a reverse J-shaped curve. This will prove crucial for the overall argument. Indeed, convexity is an exceedingly powerful mathematical condition, and permits deep inference (e.g., Rockafellar, 1970). Ellis (1985, Ch. VI) applies convexity theory to conventional statistical mechanics.

For a Gaussian channel having noise with zero mean and variance σ^2 (Cover and Thomas, 1991),

$$R(D) = 1/2\log[\sigma^2/D], 0 < D \leq \sigma^2$$

$$R(D) = 0, D > \sigma^2.$$

(2.3)

For a Poisson channel with message arrival rate λ – essentially information transmission by an analog to photon counting in which the instantaneous output rate is controlled by the input modulo noise – the rate distortion function is (Bedekar, 2001)

$$R(D) = \log[1/\lambda D], 0 < D \leq 1/\lambda$$

$$R(D) = 0, D > 1/\lambda$$

(2.4)

where D is the average in-order service time of a hypothetical first come first serve queue that would result in the output.

Lestas et al. (2010) model negative feedback in biological processes using a Poisson point process, finding that the minimum standard deviation in abundances – error – decreases only with the *quartic* root of the number of signaling events, making it energetically expensive to increase accuracy. Later we will derive similar results using a different and more general method.

These examples suggest a more general class of rate distortion functions having the form

$$R(D) = a \log[b/G_n(D)], 0 < D \leq G_n^{-1}(b),$$

$$R(D) = 0, D > G_n^{-1}(b),$$

(2.5)

where $a, b > 0$, and $G_n(D)$ is a polynomial of order $n > 0$ with at most a single zero at $D = 0$, since the rate distortion function is always a convex function of D.

Recall, now, the relation between information source uncertainty and channel capacity (e.g., Ash, 1990),$H[X] \leq C$, where H is the uncertainty of the source X and C the channel capacity, defined according to the relation (Ash, 1990)

$$C \equiv \max_{P(X)} I(X|Y),$$

(2.6)

where $P(X)$ is chosen so as to maximize the rate of information transmission along a channel Y.

Note that for a parallel set of noninteracting channels, the overall channel capacity is the sum of the individual capacities, providing a powerful 'consensus average' that does not apply in the case of modern molecular coding.

Finally, recall the analogous definition of the rate distortion function above, again an extremum over a probability distribution.

Our own work (Wallace and Wallace, 2008) focuses on the homology between information source uncertainty and free energy density. More formally, if $N(n)$ is the number of high probability 'meaningful' – that is, grammatical and syntactical – sequences of length n emitted by an information source X, then, according to the Shannon-McMillan

Theorem, the zero-error limit of the Rate Distortion Theorem (Ash, 1990; Cover and Thomas, 1991; Khinchin, 1957),

$$H[X] = \lim_{n \to \infty} \frac{\log[N(n)]}{n}$$

$$= \lim_{n \to \infty} H(X_n | X_0, ..., X_{n-1})$$

$$= \lim_{n \to \infty} \frac{H(X_0, ..., X_n)}{n+1},$$

(2.7)

where, again, $H(...|...)$ is the conditional and $H(..., ...)$ is the joint Shannon uncertainty.

In the limit of large n, $H[X]$ becomes homologous to the free energy density of a physical system at the thermodynamic limit of infinite volume. More explicitly, the free energy density of a physical system having volume V and partition function $Z(\beta)$ derived from the system's Hamiltonian – the energy function – at inverse temperature β is (e.g., Landau and Lifshitz 2007)

$$F[K] = \lim_{V \to \infty} -\frac{1}{\beta} \frac{\log[Z(\beta, V)]}{V} \equiv$$

$$\lim_{V \to \infty} \frac{\log[\hat{Z}(\beta, V)]}{V},$$

with $\hat{Z} = Z^{-1/\beta}$. The latter expression is formally similar to the first part of equation (2.7), a circumstance having deep implications: Feynman (2000) describes in great detail how information and free energy have an inherent duality. Feynman, in fact, defines information precisely as the free energy needed to erase a message. The argument is surprisingly direct (e.g., Bennett, 1988), and for very simple systems it is easy to design a small (idealized) machine that turns the information within a message directly into usable work – free energy. Information is a form of free energy and the construction and transmission of information within living things consumes metabolic free energy, with nearly inevitable losses via the second law of thermodynamics. If there

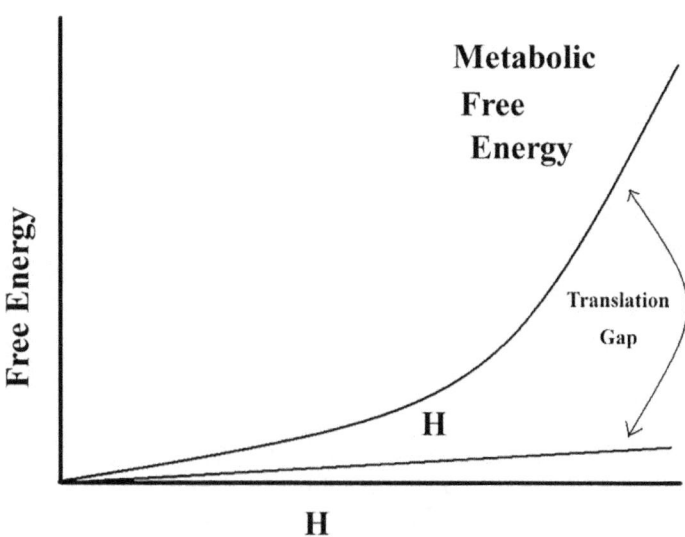

Figure 2.1: Increase in metabolic free energy needed to maintain and generate linear increase in the information source uncertainty free energy of a complex physiological process. H is seen to 'leverage' metabolic free energy expenditures, parameterizing a more complicated nonequilibrium thermodynamics. See Section 4.2.2 for a simple explicit calculation.

are limits on available metabolic free energy there will necessarily be limits on the ability of living things to process information.

Figure 2.1 presents a schematic of the mechanism: As the complexity of a dynamic physiological information process rises, that is, as H increases, its free energy content increases linearly. The metabolic free energy needed to construct and maintain the physiological systems that instantiate H should, however, be expected to increase with it at a much greater rate, hence the 'translation gap' of the figure. Section 4.2.2 below gives a fairly elementary derivation of such relations in terms of rate distortion theory. Figure 2.1 suggests that H may indeed be a good, if possibly nonlinear, index of large-scale free energy dynamics.

Conversely, information source uncertainty has an important heuristic interpretation that Ash (1990) describes as follows:

> [W]e may regard a portion of text in a particular language as being produced by an information source. The probabilities $P[X_n = a_n | X_0 = a_0, ... X_{n-1} = a_{n-1}]$ may be

estimated from the available data about the language; in this way we can estimate the uncertainty associated with the language. A large uncertainty means, by the [Shannon-McMillan Theorem], a large number of 'meaningful' sequences. Thus given two languages with uncertainties H_1 and H_2 respectively, if $H_1 > H_2$, then in the absence of noise it is easier to communicate in the first language; more can be said in the same amount of time. On the other hand, it will be easier to reconstruct a scrambled portion of text in the second language, since fewer of the possible sequences of length n are meaningful.

In sum, if a biological system characterized by H_1 has a richer and more complicated internal communication structure than one characterized by H_2, then necessarily $H_1 > H_2$ and system 1 represents a more energetic process than system 2, and by the arguments of figure 2.1, may trigger even greater metabolic free energy dynamics.

By equations (2.4-2.7), the Rate Distortion Function, $R(D)$ is likewise a free energy measure, constrained by the availability of metabolic free energy.

2.3   What 'decodes the codon'?

A number of commentators have raised the question of what actually observes the rate distortion function, i.e., what decodes the codon and makes a comparison between what is expected from the sequence of codons and the protein that is actually produced? The question is, in a different idiom, a fundamental conundrum in cellular biology, as it appears to violate the central dogma of molecular biology that information flows from DNA to RNA to protein. There is, in fact, an elaborate cellular level apparatus that does the essential decoding. Hebert and Molinari (2007) put the matter thus:

> Understanding the mechanisms regulating degradation of folding-defective polypeptides expressed in the [endoplasmic reticulum, ER] is one of the central issues of cell biology. Rapid disposal of folding-incompetent polypeptides produced in the ER lumen is instrumental to maintain ER homeostasis. The degradation machinery is easily saturated. Defective adaptation of the cellular degradation capacity to the ER load may result in accumulation of aberrant polypeptides that eventually impairs the ER capacity to assist maturation of newly synthesized secretory proteins.

Thus cells invoke a complicated, highly evolved, internal regulatory process, incorporating the elaborate machinery of the endoplasmic reticulum and chaperone proteins, that decodes the codon, i.e., compares a produced protein with an internal (inherited or learned) pattern, and then chooses one among several possible actions based on the comparison: pass the protein on to the next stage, attempt to repair a damaged protein, attempt to recycle or eliminate a protein that cannot be repaired. After further theoretical development, Chapter 3 will rephrase protein folding and regulation in terms of a cognitive paradigm that formalizes this decision process at cellular and higher – as opposed to the molecular – levels of analysis, thus finessing the apparent violation of the central dogma.

From a broader perspective, however, there is another, quite relentless if crude, mechanism for decoding the codon, and that is the continued life of the cell, tissue, or organism, i.e., Darwin's survival of the fittest, writ small: protein misfolding kills.

But the basic question – what decodes the codon – is, in fact, a misunderstanding of the regularities inherent in all information transmission. The rate distortion theorem is, to any form of information transmission, as basic as the central limit theorem is to sums of stochastic variates, and it is no surprise that many mechanisms exist in nature to decode/retranslate the codon. The essential point is that $R(D)$ is a fundamental empirical characteristic of any information transmission channel, and it can be measured by internal cellular, or external human, agencies. That is, a clever experimenter could study a cellular process, infer from the codon stream what protein should be produced, and then compare the actual with the intended product, calculating a distortion measure, and determining the minimum channel capacities needed to map out $R(D)$. This would be as arduous as, but no more arduous than, measuring the protein folding funnel, and indeed, the folding funnel and the rate distortion function appear to be different images of the same phenomenon.

The machinery of the endoplasmic reticulum and friends appears to do such measurement in situ and almost in real time, triggering heat shock protein corrective mechanisms as needed. This is no small evolutionary accomplishment.

2.4   The energy picture

Ash's comment above leads directly to a model in which the average distortion between the initial codon stream and the final form of the folded amino acid stream, the protein, becomes a dominant force, particularly in an evolutionary context in which fidelity of codon expression has survival value. The most direct model is parameterized by the average distortion between the codon stream and the folded protein structure:

Suppose there are n possible folding schemes. The most familiar approach, perhaps, is to assume that a given distortion measure, D, under evolutionary selection constraints, serves much as an external temperature bath for the possible distribution of conformation free energies, the set $\{\mathcal{H}_1, ..., \mathcal{H}_n\}$. That is, high distortion, represented by a low rate of transmission of information between codon machine and amino acid/protein folding machine, permits a larger distribution of possible symmetries – the big end of the folding funnel – according to the classic formula

$$Pr[\mathcal{H}_j] = \frac{\exp[-\mathcal{H}_j/\lambda D]}{\sum_{i=1}^{n} \exp[-\mathcal{H}_i/\lambda D]},$$

(2.8)

where $Pr[\mathcal{H}_j]$ is the probability of folding scheme j having conformational free energy $\mathcal{H}_j$.

We are, in essence, assuming that $Pr[\mathcal{H}_j]$ is a one parameter distribution in the 'intensive' quantity D.

The free energy Morse Function associated with this probability is

$$F_R = -\lambda D \log[\sum_{i=1}^{n} \exp[-\mathcal{H}_i/\lambda D]].$$

(2.9)

Applying a spontaneous symmetry breaking argument to F_R generates topological transitions in folded protein structure as the 'temperature' D decreases, i.e., as the average distortion declines. That is, as the channel capacity connecting codon machines with amino acid/protein folding machines increases, the system is driven to a particular conformation, according to the 'protein folding funnel'.

2.5   The developmental picture

The developmental approach of Wallace and Wallace (2009) permits a markedly different, and far more subtle, perspective.

We now are concerned with developmental pathways in a 'phenotype space' that, in a series of steps, take the amino acid string $\mathbf{S}_0$ at time 0 to the final folded conformation $\mathbf{S}_f$ at some time t in a long series of distinct, sequential, intermediate configurations $\mathbf{S}_i$.

Let $N(n)$ be the number of possible paths of length n that lead from $\mathbf{S}_0$ to $\mathbf{S}_f$. The essential assumptions are:

[1] This is a highly systematic process governed by a 'grammar' and 'syntax' driven by the evolutionarily-sculpted folding funnel, so that it is possible to divide all possible paths $x_n = \{\mathbf{S}_0, \mathbf{S}_1, ..., \mathbf{S}_n\}$ into two sets, a small, high probability subset that conforms to the demands of the folding funnel topology, and a much larger 'nonsense' subset having vanishingly small probability.

[2] If $N(n)$ is the number of high probability paths of length n, then the 'ergodic' limit

$$H = \lim_{n \to \infty} \frac{\log[N(n)]}{n}$$

(2.10)

both exists and is independent of the path x. This is, essentially, a restatement of the Shannon-McMillan Theorem (Khinchin, 1957).

That is, the folding of a particular protein, from its amino acid string to its final form, is not a random event, but represents a highly – evolutionarily – structured (i.e., by the folding funnel) 'statement' by an information source having source uncertainty H.

2.5.1   Symmetry arguments

A formal equivalence class algebra can now be constructed by choosing different origin and end points $\mathbf{S}_0, \mathbf{S}_f$ and defining equivalence of two states by the existence of a high probability meaningful path connecting them with the same origin and end. Disjoint partition by equivalence class, analogous to orbit equivalence classes for dynamical systems, defines the vertices of the proposed network of developmental protein 'languages'. We thus envision a *network of metanetworks*. Each vertex then represents a different equivalence class of developmental information sources. This is an abstract set of metanetwork 'languages'.

This structure generates a groupoid, in the sense of the Appendix. States a_j, a_k in a set A are related by the groupoid morphism if and only if there exists a high probability grammatical path connecting them to the same base and end points, and tuning across the various

possible ways in which that can happen – the different developmental
languages – parameterizes the set of equivalence relations and creates
the (very large) groupoid.

There is an implicit hierarchy. First, there is structure *within the
system having the same base and end points*. Second, there is a complicated groupoid structure defined by sets of dual information sources
surrounding the variation of base and end points. We do not need
to know what that structure is in any detail, but can show that its
existence has profound implications.

We begin with the simple case, the set of dual information sources
associated with a fixed pair of beginning and end states.

The first level

Taking the serial grammar/syntax model above, we find that not all
high probability meaningful paths from S_0 to S_f are actually the same.
They are structured by the uncertainty of the associated dual information source, and that has a homological relation with free energy
density.

Let us index possible information sources connecting base and end
points by some set $A = \cup\alpha$. Argument by abduction from statistical
physics is direct. The minimum channel capacity needed to produce
average distortion less than D in the energy picture above is $R(D)$. We
take the probability of a particular H_α as determined by the standard
expression

$$P[H_\beta] = \frac{\exp[-H_\beta/\mu R]}{\sum_\alpha \exp[-H_\alpha/\mu R]},$$

(2.11)

where the sum may, in fact, be a complicated abstract integral.

A basic requirement, then, is that the sum/integral always converges.

Thus, in this formulation, there must be structure *within* a (cross
sectional) connected component in the base configuration space, determined by R. Some dual information sources will be 'richer'/smarter
than others, but, conversely, must use more available channel capacity
for their completion.

The second level

While we might simply impose an equivalence class structure based on equal levels of energy/source uncertainty, producing a groupoid – and possibly allowing a Morse Theory approach – we can do more *by now allowing both source and end points to vary*, as well as by imposing energy-level equivalence. This produces a far more highly structured groupoid.

Equivalence classes define groupoids, by standard mechanisms (Weinstein, 1996), as described in the Appendix. The basic equivalence classes – here involving both information source uncertainty level and the variation of $\mathbf{S}_0$ and $\mathbf{S}_f$, will define transitive groupoids, and higher order systems can be constructed by the union of transitive groupoids, having larger alphabets that allow more complicated statements in the sense of Ash above.

Again, given a minimum necessary channel capacity R, we propose that the metabolic-energy-constrained probability of an information source representing equivalence class G_i, H_{G_i}, will again be given by

$$P[H_{G_i}] = \frac{\exp[-H_{G_i}/\kappa R]}{\sum_j \exp[-H_{G_j}/\kappa R]},$$

(2.12)

where the sum/integral is over all possible elements of the largest available symmetry groupoid. By the arguments of Ash above, compound sources, formed by the union of underlying transitive groupoids, being more complex, generally having richer alphabets, as it were, will all have higher free-energy-density-equivalents than those of the base (transitive) groupoids.

Let

$$Z_G = \sum_j \exp[-H_{G_j}/\kappa R].$$

(2.13)

We now define the *Groupoid free energy* of the system, a Morse Function F_G, at channel capacity R, as

$$F_G[R] = -\kappa R \log[Z_G[R]].$$

(2.14)

These free energy constructs permit introduction of the spontaneous symmetry breaking arguments above, but now an *increase* in R (with corresponding decrease in average distortion D) permits richer system dynamics – higher source uncertainty – resulting in more rapid transmission of the 'message' constituting convergence from $\mathbf{S}_0$ to $\mathbf{S}_f$.

2.5.2   Folding speed and mechanism

Dill et al. (2007) describe the conundrum of folding speeds as follows:

> ...[P]rotein folding speeds – now known to vary over more than eight orders of magnitude – correlate with the topology of the native protein: fast folders usually have mostly local structure, such as helices and tight turns, whereas slow folders usually have more non-local structure, such as β sheets (Plaxco et al., 1998)...

A simple rate distortion argument reproduces this result. Assume that protein structure can be characterized by some groupoid representing, at least, the disjoint union of the groups describing the symmetries of component secondary structures – e.g., helices and sheets. Then, in equation (2.11), the set $A = \cup \alpha$ grows in size – cardinality – with increasing structural complexity. If channel capacity is capped by some mechanism, so that (at least) R grows at a lesser rate than A, by some measure, then

$$P[H_\beta] = \frac{\exp[-H_\beta/\mu R]}{\sum_\alpha \exp[-H_\alpha/\mu R]}$$

(2.15)

must decrease with increase in the number of possible states α, i.e., with increase in the cardinality of A, producing progressively lower rates of convergence to the final state.

In particular, if R is fixed, then the log of the folding rate will be given as

$$\log[P[H_\beta]] = \log[\frac{\exp[-H_\beta/\mu R]}{\sum_\alpha \exp[-H_\alpha/\mu R]}] \equiv C(R) - H_\beta/\mu R,$$

(2.16)

where $C(R)$ is positive. β indexes increasing topological complexity, using some appropriate measure.

For simplicity, assume $H_\beta \propto \beta$. Then, taking an integral approximation,

$$P[\beta] \approx \frac{\exp[-m\beta/\mu R]}{\int_{\alpha=0}^{\infty} \exp[-m\alpha/\mu R]d\alpha} = (m/\mu R)\exp[-m\beta/\mu R],$$

(2.17)

and

$$\log[P[\beta]] \approx \log[m/\mu R] - m\beta/\mu R.$$

(2.18)

Thus one expects, at a fixed R defining a maximum channel capacity, that

$$\log[FoldingRate] \approx C - k\beta,$$

(2.19)

C, k constant and all values positive.

A standard empirical index of protein complexity is the absolute contact order (Plaxco et al, 1998):

$$ACO = 1/N \sum_{}^{N} \Delta L_{i,j}$$

(2.20)

where N is the number of contacts within 6 Angstroms between non-hydrogen atoms in the protein, and $\Delta L_{i,j}$ is the number of residues separating the interacting pair of nonhydrogen atoms.

Adjacent residues are assumed to be separated by one residue.

Figure 2.2, adapted from Gruebele (2005), shows the correlation of the log of the folding rate with fold complexity, measured by the ACO. The upper line estimates folding speed limited only by fold complexity, following Yang and Gruebele (2004), and seems clearly to represent a maximum possible rate distortion function/channel capacity, according to equation (2.19). The molecular species along the lower curve are assumed to be 'frustrated' by an irregular folding funnel, and follow a narrow spectrum of relations like equation (2.19), necessarily below the line defined by maximum channel capacity, and necessarily somewhat scattered, according to the variation in R.

It is possible to generate something like figure 2.2 by describing 'smooth' and 'rough' folding funnels in terms of a Gaussian, Poisson, or other channel in which the signal transmission from initial to final protein state is perturbed by noise that parameterizes the rate distortion function. For example, plugging equations (2.3) or (2.4) – Gaussian or Poisson channels – into equation (2.18) gives, over an appropriate range of parameters, a spectrum of linear relations for log folding rate like that in figure 2.3. D, m, and μ are fixed, and β, λ, and σ^2 increase as indicated. Equation (2.18), however, is quite general, whatever the parameterization of $R(D)$ by the roughness of the folding funnel – e.g., by λ and σ^2 – so that the result is robust.

These matters lead to the next central question: can folding rates be modulated by other means than noise in the folding funnel? Can the effects of noise be 'reversed'? This will lead toward our cognitive model for protein folding.

2.5.3   Catalysis of protein folding

Incorporating the influence of embedding contexts – epigenetic or cellular regulatory chaperone effects, or the impact of (broadly) toxic expo-

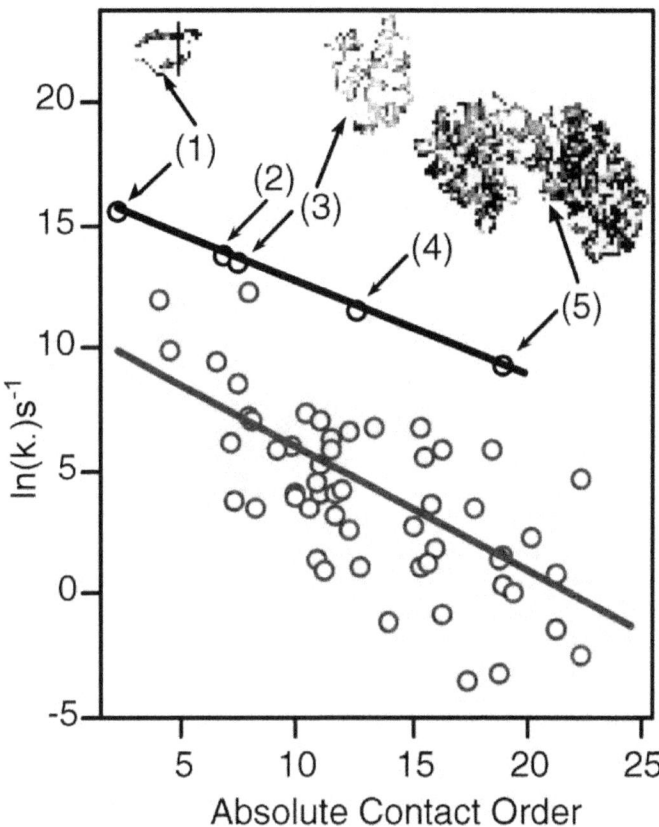

Figure 2.2: From Gruebele, (2005). Relation of the log of the protein folding rate to fold complexity. The upper folding speeds are limited only by fold complexity, without the 'frustration' effects of a rough folding funnel. Frustration, in this model, is equivalent to increasing noise that constrains channel capacity, and drives R irregularly lower than the value implied by the relation for the fastest folders. Equations (2.18) and (2.19) reproduce something of these results.

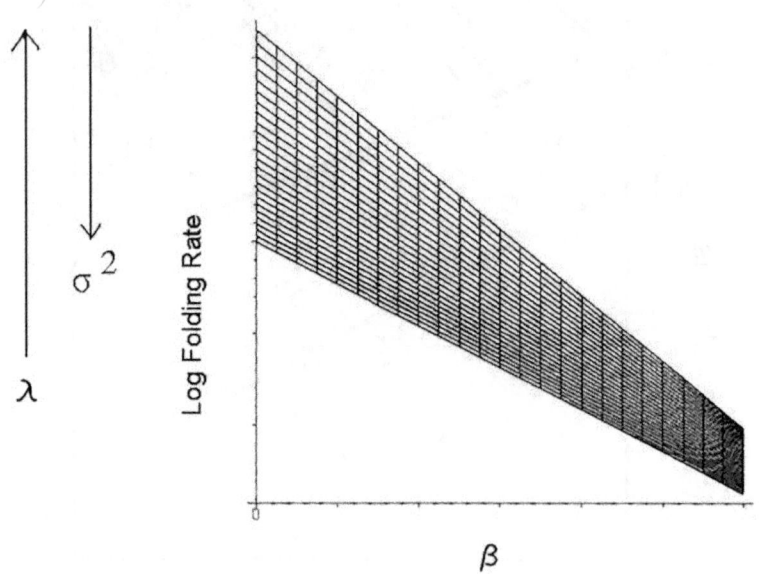

Figure 2.3: Spectrum of linear relations between log folding rate and increasing topological complexity for increasing 'roughness' of the folding funnel, as measured by noise σ^2 for a Gaussian channel or 'photon rate' λ for a Poisson channel. β increases to the right, σ^2 increases downward, and λ upward.

sures – can be done here by invoking the Joint Asymptotic Equipartition Theorem (JAEPT)(Cover and Thomas, 1991). For example, given an embedding contextual information source, say Z, that affects protein development, then the developmental source uncertainty $H_{X_{G_i}}$ is replaced by a joint uncertainty $H(X_{G_i}, Z)$. The objects of interest then become the jointly typical dual sequences $y^n = (x^n, z^n)$, where x is associated with protein folding development and z with the embedding context. Restricting consideration of x and z to those sequences that are in fact jointly typical allows use of the information transmitted from Z to X as the splitting criterion.

One important inference is that, from the information theory 'chain rule' (Cover and Thomas, 1991),

$$H(X, Z) = H(X) + H(Z|X) \leq H(X) + H(Z),$$

(2.21)

while there are approximately $\exp[nH(X)]$ typical X sequences, and $\exp[nH(Z)]$ typical Z sequences, and hence $\exp[n(H(X) + H(Z))]$ independent joint sequences, there are only about

$$\exp[nH(X, Z)] \leq \exp[n(H(X) + H(Z))]$$

jointly typical sequences, so that the effect of the embedding context, in this model, is to lower the *relative* free energy of a particular protein channel.

Thus the effect of epigenetic/catalytic regulation or toxic exposure is to channel protein folding into pathways that might otherwise be inhibited or slowed by an energy barrier. Hence the epigenetic/catalytic/toxic information source Z acts as a *tunable catalyst*, a kind of second order enzyme, to enable and direct developmental pathways. This result permits hierarchical models similar to those of higher order cognitive neural function (e.g, Wallace, 2005).

This is indeed a relative energy argument, since, metabolically, two systems must now be supported, i.e., that of the 'reaction' itself and that of its catalytic regulator. 'Programming' and stabilizing inevitably intertwined, as it were.

Protein folding, in the developmental picture, can be visualized as a series of branching pathways. Each branch point is a developmental decision, or switch point, governed by some regulatory apparatus (if only the slope of the folding funnel) that may include the effects of toxins or epigenetic mechanisms.

A more general picture emerges by allowing a distribution of possible 'final' states $\mathbf{S}_f$. Then the groupoid arguments merely expand to permit traverse of both initial states and possible final sets, recognizing that there can now be a possible overlap in the latter, and the catalytic effects are realized through the joint uncertainties $H(X_{G_i}, Z)$, so that the guiding information source Z serves to direct as well the possible final states of X_{G_i}.

2.5.4   Generalizing the developmental model

The most natural extension of the developmental model of protein folding would be in terms of the directed homotopy classification of ontological trajectories, in the sense of Wallace and Wallace (2008, 2009). That is, developmental trajectories themselves can be classified into equivalence classes, for example those that lead to a normal final state $\mathbf{S}_f$, and those that lead to pathological aggregations or misfoldings, say some set $\{\mathbf{S}_{path}^i\}, i = 1, 2, \dots.$ This produces a dynamic directed homotopy groupoid topology whose understanding might be useful across a broad spectrum of diseases.

Figure 2.4 illustrates the concept. The initial developmental state $\mathbf{S}_0$ can, in this picture, 'fall' down two different sets of developmental pathways, separated by a critical period 'shadow' preventing crossover between them. Paths within one set can be topologically transformed into each other without crossing the filled triangle, and constitute a directed homotopy equivalence classes. The lower apex of the triangle can, however, start at many possible critical period points along any path connecting $\mathbf{S}_0$ and $\mathbf{S}_f$, following the arguments of Section 12 of Wallace and Wallace (2009).

Onset of a path that converges on the conformation $\mathbf{S}_{path}$ is, according to the model, driven by a genetic, epigenetic, or environmental catalysis event, in the sense of Section 2.5.3. The topological equivalence classes define a groupoid on the developmental system.

Some further argument in this direction is possible, invoking the metabolic costs suggested by figure 2.1.

Of necessity, from equation (2.21),

$$H(X, Z) < H(X) + H(Z)$$

(2.22)

if $H(Z|X) < H(Z)$.

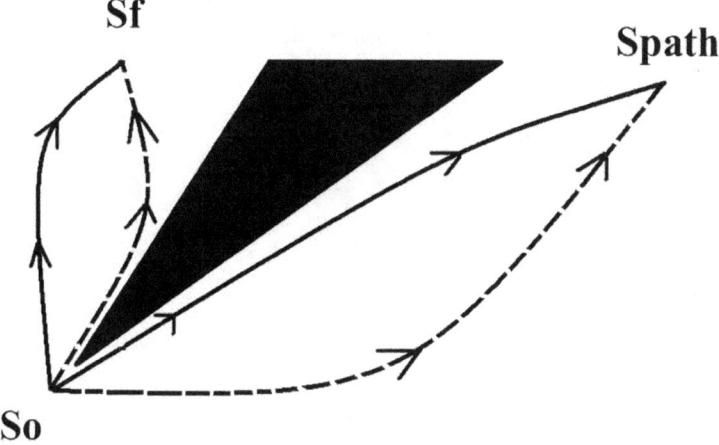

Figure 2.4: Given an initial state S_0 and a critical period casting a path-dependent developmental shadow, there are two different directed homotopy equivalence classes of deformable paths leading, respectively, to the normal folded protein state S_f and the pathological state – e.g., amyloid – S_{path}. These sets of paths form equivalence classes defining a topological groupoid.

These relations imply that, by means of the identification of information as a form of free energy, at the expense of adding the considerable energy burden of some regulatory apparatus, represented by its dual information source Z, it becomes possible to canalize the reaction paths of the developmental picture, so as to make one set of pathways beginning with S_0 far more probable than any other.

That is, by raising the entire reaction free energy landscape corresponding to $H(X)$ by the amount $H(Z)$ it becomes possible to deepen the energy channel leading from S_0 to the desired outcome, S_f. Complicated internal reaction mechanisms can be subsumed by the Shannon-McMillan Theorem, in the same sense that the Central Limit Theorem subsumes the behavior of long sums of stochastic variates into the Normal distribution.

Within a cell or organism, however, there will be always an ensemble of possible developmental states and pathways, driven by available metabolic free energy, so that, taking $< .. >$ as representing an average,

$$[< H(X, Z) >] < [< H(X) > + < H(Z) >].$$

(2.23)

Typically, letting M represent the intensity of available metabolic free energy, a rate index, one expects

$$< H > \approx \frac{\int H \exp[-H/\kappa M] dH}{\int \exp[-H/\kappa M] dH} \approx \kappa M,$$

(2.24)

where κ, an inverse energy rate scaling constant, may be quite small indeed, a consequence of entropic translation losses between metabolic free energy and the expression of information.

The resulting relation,

$$M_{X,Z} < M_X + M_Z,$$

(2.25)

suggests an explicit free energy mechanism for developmental canalization.

If entropic translation losses are not linear with increase in information transmission rate H, we might replace κM with some function $Q(\kappa M)$ that 'tops out' with increasing M, for example $Q \propto \log[\kappa M]$. This means that, after a certain point, large increases in metabolic free energy are needed to increase biological information. The energy relation then becomes, after a little algebra,

$$M_{X,Z} < \kappa \times M_X \times M_Z \ll M_X + M_Z,$$

(2.26)

if either κ or one of the other M-terms is small, and a low energy information source regulator could thus be used to 'leverage' reaction canalization very efficiently.

In reality, there will be a large, nested, set of appropriately coarse-grained regulatory and/or signaling processes expressed as information sources – $Y_1, ..., Y_m$ – in which the physiological system of interest will be embedded, ranging from other physiological systems to patterns of social and cultural interaction, mediated by epigenetic inheritance across generations. The relations of interest then become

$$[< H(X, Y_1, ..., Y_m) >] < [< H(X) > + \sum_j < H(Y_j) >],$$

$$M_{X,Y_1,...,Y_m} < M_X + \sum_j M_{Y_j},$$

$$M_{X,Y_1,...,Y_m} < \kappa^m M_X \Pi_j M_{Y_j} \ll M_X + \sum_j M_{Y_j}.$$

(2.27)

That is, quite counterintuitively, entropic loss can be a powerful tool for triggering complex biological logic gates like figure 2.4, in much the same sense that Tompa and Csermely (2004) propose that entropy transfer can be used by generalized chaperones to trigger proper conformation in pathologically folded protein complexes.

The last expression above suggests, in particular, that the appropriate regulatory level for intervention *may not be that of the desired target*. That is, the model implies that the most efficient intervention may be upstream from the desired target or, more likely, involve synergistic dynamic intrusions at more than one scale or level of organization to bring down the overall magnitude of the product term.

This is an essential result that illustrates the ultimate necessity of multilevel/multiscale therapeutic interventions: effective magic bullet treatments are likely to prove relatively rare in the biological thickets of the real world.

Chapter 3

The cognitive paradigm

3.1 The basic idea

We now take the developmental perspective as the foundation for generating an empirically-based statistical model – effectively a cognitive paradigm for normal and pathological protein folding – that incorporates the embedding contexts of epigenetic and environmental signals. Atlan and Cohen (1998), in their study of the immune system, argue that the essence of cognition is the comparison of a perceived signal with an internal, learned picture of the world, and then choice of a single response from a large repertoire of possible responses. Such choice inherently involves information and information transmission since it always generates a reduction in uncertainty, as explained in Ash (1990, p. 21). Thus cognitive structures, by processing information, are constrained by the asymptotic limit theorems of information theory, in the same sense that sums of stochastic variables are constrained by the Central Limit Theorem, allowing the construction of powerful statistical tools useful for data analysis.

More formally, a pattern of incoming input $\mathbf{S}_i$ describing the folding status of the protein – starting with the initial codon stream $\mathbf{S}_0$ – is mixed in a systematic algorithmic manner with a pattern of otherwise unspecified 'ongoing activity', including cellular, epigenetic and environmental signals, $\mathbf{W}_i$, to create a path of combined signals $x = (a_0, a_1, ..., a_n, ...)$. Each a_k thus represents some functional composition of internal and external factors, and is expressed in terms of the intermediate states as

$$\mathbf{S}_{i+1} = f([\mathbf{S}_i, \mathbf{W}_i]) = f(a_i)$$

41

(3.1)

for some unspecified function f. The a_i are seen to be very complicated composite objects, in this treatment that we may choose to coarse-grain so as to obtain an appropriate 'alphabet'.

In a simple spinglass-like model, $\mathbf{S}$ would be a vector, $\mathbf{W}$ a matrix, and f would be a function of their product at 'time' i.

The path x is fed into a highly nonlinear decision oscillator, h, a 'sudden threshold machine' pattern recognition structure, in a sense, that generates an output $h(x)$ that is an element of one of two disjoint sets B_0 and B_1 of possible system responses. Let us define the sets B_k as

$$B_0 = \{b_0, ..., b_k\},$$

$$B_1 = \{b_{k+1}, ..., b_m\}.$$

Assume a graded response, supposing that if $h(x) \in B_0$, the pattern is not recognized, and if $h(x) \in B_1$, the pattern has been recognized, and some action $b_j, k+1 \le j \le m$ takes place. Typically, the set B_1 would represent the final state of the folded protein, either normal or in some pathological conformation, that is sent on in the biological process or else subjected to some attempted corrective action. Corrections may, for example, range from activation of 'heat shock' protein repair to more drastic clean-up attack.

The principal objects of formal interest are paths x triggering pattern recognition-and-response. That is, given a fixed initial state $a_0 = [\mathbf{S}_0, \mathbf{W}_0]$, examine all possible subsequent paths x beginning with a_0 and leading to the event $h(x) \in B_1$. Thus $h(a_0, ..., a_j) \in B_0$ for all $0 < j < m$, but $h(a_0, ..., a_m) \in B_1$. B_1 is thus the set of final possible states, $\{\mathbf{S}_f\} \cup \{\mathbf{S}_{path}\}$ from figure 2.4 that includes both the final 'physics' state $\mathbf{S}_f$ and the set of possible pathological conformations.

Again, for each positive integer n, let $N(n)$ be the number of high probability grammatical and syntactical paths of length n which begin with some particular a_0 and lead to the condition $h(x) \in B_1$. Call such paths 'meaningful', assuming, not unreasonably, that $N(n)$ will be considerably less than the number of all possible paths of length n leading from a_0 to the condition $h(x) \in B_1$.

While the combining algorithm, the form of the nonlinear oscillator, and the details of grammar and syntax, can all be unspecified in this model, the critical assumption that permits inference of the necessary conditions constrained by the asymptotic limit theorems of information theory is that the finite limit

$$H = \lim_{n \to \infty} \frac{\log[N(n)]}{n}$$

(3.2)

both exists and is independent of the path x.

Call such a pattern recognition-and-response cognitive process *ergodic*. Not all cognitive processes are likely to be ergodic in this sense, implying that H, if it indeed exists at all, is path dependent, although extension to nearly ergodic processes seems possible (e.g., Wallace and Fullilove, 2008, pp. 27-28).

Invoking the spirit of the Shannon-McMillan Theorem, as choice involves an inherent reduction in uncertainty, it is then possible to define an adiabatically, piecewise stationary, ergodic (APSE) information source $\mathbf{X}$ associated with stochastic variates X_j having joint and conditional probabilities $P(a_0, ..., a_n)$ and $P(a_n | a_0, ..., a_{n-1})$ such that appropriate conditional and joint Shannon uncertainties satisfy the classic relations of equation (2.7).

This information source is defined as *dual* to the underlying ergodic cognitive process.

Adiabatic means that the source has been parameterized according to some scheme, and that, over a certain range, along a particular piece, as the parameters vary, the source remains as close to stationary and ergodic as needed for information theory's central theorems to apply. *Stationary* means that the system's probabilities do not change in time, and *ergodic*, roughly, that the cross sectional means approximate long-time averages. Between pieces it is necessary to invoke various kinds of phase transition formalisms, as described more fully in e.g., Wallace (2005).

Structure is now subsumed *within the sequential grammar and syntax of the dual information source* rather than within the set of developmental paths of figure 2.4 and the added catalysis arguments of Section 2.5.4.

This transformation in perspective carries heavy computational burdens, as well as providing deeper mathematical insight, as cellular machineries, and phenomena of epigenetic or environmental catalysis, are now included within a single model.

3.2   Coevolution models

The most evident assumption at this point is that there may be more than a single cognitive protein folding process in operation, e.g., that the action of the endoplasmic reticulum, chaperones, and other corrective mechanisms, involves separate cognitive chemical phenomena $\{H_1, ..., H_m\}$ that interact via some form of crosstalk. Following the direction of Wallace and Wallace (2009) we invoke an internal system of 'empirical Onsager relations', assuming that the different cognitive processes represented by these dual information sources *become each others primary environments*, a broadly, if locally, coevolutionary phenomenon, in the sense of Diekmann and Law (1996). We write

$$H_k = H_k(K_1, ..., K_s, ..., H_j, ...)$$

(3.3)

where the K_s represent other relevant parameters and $k \neq j$. In a generalization of the statistical model, we would expect the dynamics of such a system to be driven by an empirical recursive network of stochastic differential equations. Letting the K_s and H_j all be represented as parameters Q_j, with the caveat that H_k not depend on itself, we are able to define an entropy-analog based on the homology of information source uncertainty with free energy as

$$S_k = H_k - \sum_i Q_i \partial H_k / \partial Q_i,$$

(3.4)

whose gradients in the Q, having the form $\partial S_k / \partial Q_j$, define local (broadly) chemical forces. Most simply, for such systems, (e.g., deGroot and Mazur, 1984) we assume a set of linear regression-like dynamic relations,

$$dQ_i / dt = \sum_{j,k} \mu^i_{j,k} \partial S_j / \partial Q_k,$$

where the μ are constant. Note that, since information systems are not locally time-reversible, there can be no 'reciprocity' in indices j, k.

Ultimately, in close analogy with nonequilibrium thermodynamics, this generalizes to a recursive system of phenomenological Onsager-like stochastic differential equations(Wallace and Wallace, 2008, 2009):

$$dQ_t^j = \sum_i [L_{j,i}(t, ..., Q_k, ...)dt + \sigma_{j,i}(t, ..., Q_k, ...)dB_t^i]$$

(3.5)

where, again, for notational simplicity, we have expressed both parameters and information sources in terms of the same symbols Q^k. The dB_t^i represent different kinds of 'noise' having particular forms of quadratic variation that may represent a projection of environmental factors under something like a rate distortion manifold (Glazebrook and Wallace, 2009a, b). See the Mathematical Appendix for a brief introduction to stochastic differential equations.

There are several obvious possible dynamic patterns for this system of equations:

1. Setting equation (3.5) equal to zero and solving for stationary points gives attractor states since the noise terms preclude unstable equilibria.

2. The system may converge to limit cycle or pseudorandom 'strange attractor' behaviors in which it seems to chase its tail endlessly within a limited venue – a traditional coevolutionary 'Red Queen' (Wallace and Wallace, 2009).

3. What is converged to in both cases is not a simple state or limit cycle of states. Rather it is an equivalence class, or set of them, of highly dynamic information sources coupled by mutual interaction through crosstalk. Thus 'stability' in this structure represents particular patterns of ongoing dynamics rather than some identifiable static configuration.

Here we become deeply enmeshed in a system of highly recursive phenomenological stochastic differential equations, but in a dynamic rather than static manner. The objects of interest are equivalence classes of information sources, rather than simple 'stationary states'. Imposition of necessary conditions from the asymptotic limit theorems of communication theory has beaten the mathematical thicket back one full layer.

These results are essentially similar to those of Diekmann and Law (1996), who invoke evolutionary game dynamics to obtain a first order canonical equation for coevolutionary systems having the form

$$ds_i/dt = K_i(s)\partial W_i(s_i', s)|_{s_i'=s_i}.$$

(3.6)

The s_i, with $i = 1, ..., N$, denote adaptive trait values in a community comprising N species. The $W_i(s_i', s)$ are measures of fitness of individuals with trait values s_i' in the environment determined by the resident trait values s, and the $K_i(s)$ are non-negative coefficients, possibly distinct for each species, that scale the rate of evolutionary change. Adaptive dynamics of this kind have frequently been postulated, based either on the notion of a hill-climbing process on an adaptive landscape or some other sort of plausibility argument.

When this equation is set equal to zero, so there is no time dependence, one obtains what are characterized as 'evolutionary singularities' or stationary points.

Equation (3.5) above is similar, although focused on information sources representing protein folding regulation, allowing elaborate patterns of phase transition punctuation in a natural manner.

Champagnat et al. (2006), in fact, derive a higher order canonical approximation extending equation (3.6) that is closer to equation (3.5), i.e., a stochastic differential equation describing coevolutionary dynamics. Champagnat et al. extend the argument, using a large deviations approach to analyze dynamical coevolutionary paths, not merely quasi-stable singularities. They contend that in general, the issue of evolutionary dynamics drifting away from trajectories predicted by the canonical equation can be investigated by considering the asymptotic of the probability of 'rare events' for the sample paths of the diffusion.

By 'rare events' they mean diffusion paths drifting far away from the canonical equation. The probability of such rare events is governed by a large deviation principle: when a critical parameter (designated ϵ) goes to zero, the probability that the sample path of the diffusion is close to a given rare path ϕ decreases exponentially to 0 with rate $I(\phi)$, where the 'rate function' I can be expressed in terms of the parameters of the diffusion. This allows study of long-time behavior of the diffusion process when there are multiple attractive singularities. Under proper conditions the most likely path followed by the diffusion when exiting a basin of attraction is the one minimizing the rate function I over all the appropriate trajectories. The time needed to exit the basin is of the order $\exp(H/\epsilon)$ where H is a quasi-potential representing the minimum of the rate function I over all possible trajectories.

An essential fact of large deviations theory is that the rate function I which Champagnat et al. invoke can almost always be expressed as a kind of entropy, that is, in the form $I = -\sum_j P_j \log(P_j)$ for some probability distribution. This result goes under a number of names; Sanov's Theorem, Cramer's Theorem, the Gartner-Ellis Theorem, the Shannon-McMillan Theorem, and so forth (e.g., Dembo and Zeitouni, 1998). A detailed example is given in R. Wallace and R.G. Wallace (2008).

These considerations lead very much in the direction of equation (3.5), seen as subject to internally-driven large deviations *that are themselves described as information sources*, providing H-parameters that can trigger punctuated shifts between quasi-stable modes, in addition to resilience transitions driven by external catalytic events.

Indeed, the direct inclusion of large deviations regularities within the context of the statistical model of equation (3.5) suggests that other factors that can be characterized in terms of information sources may be directly included within the formalism. Section 6.1 of Wallace et al. (2009), for example, explores the impact of culture, taken as a generalized language, on the evolution of human pathogens.

The basic statistical model is illustrated by figure 3.1. Here, two quasi-equilibria – one normal, one pathological – are characterized by diffusive drift about their singularities in a two dimensional system, but are coupled by a highly structured large deviation connecting them. That large deviation excursion is by means of an information source having a 'grammar' and a 'syntax', rather than representing a random event. Understanding that grammar and syntax would, in this model, represent understanding the etiology of a protein folding disorder, a matter treated in Chapter 5 in terms of structured psychosocial stress affecting an HPA axis 'tuned' to that stress.

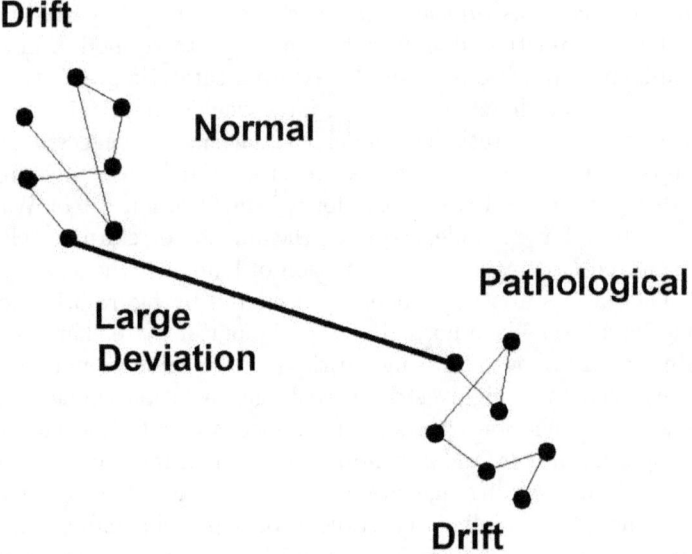

Figure 3.1: Adapted from Wallace (2010c). Dynamic behavior of the system obtained by setting equation (3.5) to zero. Diffusive drift about a 'normal' protein folding quasi-equilibrium is interrupted by a highly structured large deviation, characterized by an information source, leading to a pathological quasi-equilibrium. Chapter 5 examines such transitions as driven by psychosocial and other stress.

Chapter 4

Extending the model

4.1 Intrinsically disordered proteins I

To this point we have considered 'traditional' processes of protein folding, i.e., from codon to tertiary protein structure. As Serdyuk (2007) discusses, many proteins in the cell have no unique tertiary structure in isolation, although they have a distinct function under physiological conditions, that is, in partnership, and are implicated in a spectrum of disorders including cancer, cardiovascular disease, amyloidoses, neurodegenerative diseases and diabetes (e.g., Uversky et al., 2008). Essentially, their conformation is determined not only by their amino acid sequence, but also by the interacting partner. Basically, they are without a hydrophobic core.

Serdyuk (2007), in fact, proposes four fundamental protein forms:

> In addition to the axiomatic native state – a rigid tertiary structure – it is proposed to consider three more states: molten globule, completely disordered chain, and the structure comprising domains connected with long enough linkers and containing, as a rule, rather long disordered regions at the ends. These four states cover all conformations that are now known based on the physical understanding of protein structure.

Thus the spectrum of figure 1.1 becomes an expansion, via subgroupoids, of a single class within this larger taxonomy, having the magic number four, implying a larger, embedding, 'spherical' error code topology, in Tlusty's sense.

It is possible, from the results of Chapter 2, to say something about the rates of binding of intrinsically disordered proteins (IDP) to their targets. According to Huang and Liu (2009), these rates are greater than for ordered proteins, via a highly flexible 'fly-casting' mechanism, in contrast with ordered proteins that must dock to their targets. IDPs have, then, greater effective capture radius and can weakly bind to targets from a larger distance, and then 'reel themselves in' to the final configuration. Huang and Liu conclude that both fractions of the native interchain contacts and the distance between mass centers, quantities widely used in protein binding problems, only partially describe the features of the binding process, so that better coordinates will be required.

Taking the perspective of Section 2.5.2, we assume that IDPs are transferred hand-to-hand, to avoid cellular clean-up processes. Thus, for a given IDP, symbolized by I, there is an initial partner, i, and a final partner f, defining larger-scale structures. Then we are interested in the number of possible paths, $N(n)$, having n steps leading from an initial partnership $\mathbf{S}_{i \star I}$ to a final partnership $\mathbf{S}_{f \star I}$. This instantiates a larger version of the metanetwork of 'languages' discussed above.

Collapsing the argument via equations (2.16) and (2.19), the 'fly-casting' mechanism might better be described as a snake slithering down a rocky hillside. That is, an IDP 'falling' down a noisy folding funnel undergoes a self-lubricating catalysis that decreases σ^2 (or increases λ, depending on the nature of the channel) in something like figure 2.3, increasing the rate of reaction above what would be expected from a rigid molecule having fixed tertiary structures. Indeed, the child's toy slinky-spring walking down a staircase comes to mind, and is probably not a bad model.

Reiterating the argument from another viewpoint, Liu et al. (2009) find that 'increasing disorder' of an IDP can tune its binding affinity so as to maximize the specificity of promiscuous interactions. Assuming that the final bound state of the IDP and its target partnership, that we have written as $\mathbf{S}_{f \star I}$, can be characterized by an index of topological complexity like the absolute contact order of figure 2.2, then fixing β in figure 2.3, the 'increasing disorder' of the IDP is, in fact, an index of its propensity for self-lubrication, i.e., its inherent 'snakeiness', measured by decreasing σ^2 or increasing λ for Gaussian or Poisson folding funnel channels, respectively.

Remember that the exact nature of the folding funnel channel is not particularly important in this model, as the rate distortion function $R(D)$ is always a convex – reverse J-shaped – function of D, and this will always produce a spectrum of linear relation like those of figure 2.3 that will be parameterized by some snakiness index representing the 'increasing disorder' of the IDP.

Exact topology aside, for the moment, a very general argument provides some insight. IDP's convey information, that is, they signal more elaborate cellular machinery, and the channel capacity of that conveyance determines the maximum rate at which information can be sent, i.e., the maximum possible reaction rate. The simplest possible model is the Gaussian channel, assuming a real-number coarse-graining, acting under the 'power constraint' $E(X^2) \leq \mathcal{P}$, where X is the stochastic reaction variable of interest. If σ^2 is the variance of the (zero-mean) noise, then the channel capacity, the maximum rate at which information can be transmitted, the maximum 'reaction rate' in a real-number coarse-graining, is given by (Cover and Thomas, 1991)

$$C = \frac{1}{2}\log[1 + \frac{\mathcal{P}}{\sigma^2}].$$

In a physiological context, $\mathcal{P}$ may be a function of available metabolic free energy.

The self-lubricating snakiness argument implies that molecular flexibility decreases σ^2 in a rough folding funnel, increasing reaction rate.

These considerations suggest the snake model may have some actual use.

4.2 Aging and protein folding

4.2.1 Rate distortion Onsager models

The developmental perspective of Chapter 2, although focused on the relatively short time frames of protein metabolism – in the range from microseconds to minutes – is suggestive. The principal 'risk factor' for a large array of protein folding disorders is biological age – for humans, in the range of decades – and a simplified version of the previous chapter may provide a life-course perspective, that is, a developmental model over a far longer timescale.

Equations (2.2-2.7) suggest that the rate distortion function, $R(D)$, is itself a free energy measure, as it represents the minimum channel capacity needed to assure average distortion equal to or less than D. Let us now consider the principal branch in figure 2.4, the set of paths from $\mathbf{S}_0$ to $\mathbf{S}_f$, representing normal protein folding, taken as a communication channel having a given rate distortion function. The arguments of the previous chapter suggest that there will be an empirical Onsager relation in the gradient of the *rate distortion disorder*, an entropy-analog,

$$S_R \equiv R(D) - DdR(D)/dD$$

(4.1)

such that, over a life-history timeline,

$$dD/dt = g(dS_R/dD)$$

(4.2)

for some appropriate function g.

For Gaussian and Poisson channels, having

$$R(D) = (1/2)\log(\sigma^2/D),\ S_R(D) = (1/2)\log(\sigma^2/D) + 1/2$$

and

$$R(D) = \log(1/\lambda D),\ S_R(D) = \log(1/\lambda D) + 1,$$

respectively, the simplest possible Onsager relations become

$$dD/dt = -\mu dS_R dD = \mu/2D, (G)$$

$$dD/dt = -\mu dS_R dD = \mu/D, (P),$$

(4.3)

with the explicit solutions

$$D_G = \sqrt{\mu t},$$

$$D_P = \sqrt{2\mu t}.$$

(4.4)

For the Gaussian channel on an appropriate timescale – necessarily many orders of magnitude longer than the time of folding itself – the average distortion, representing the degree of misfolding, simply grows as a diffusion process in time. This is the simplest possible aging model, in which μ represents the accumulated impacts of epigenetic and broadly environmental effects including toxic exposures, nutrition, the richness of social interaction, and so on, over a lifetime. The Poisson channel the average distortion – a delay measure – grows at a rate $\sqrt{2}$ times greater than for a simple diffusion model.

A somewhat less simplistic model takes the Onsager relation as constrained by the availability of metabolic free energy, M, that powers active chaperone processes, respectively for Gaussian and Poisson channels,

$$dD/dt = -\mu dS_{RG}/dD - f(\kappa M) = \mu/2D - f(\kappa M),$$

$$dD/dt = -\mu dS_{RP}dD - f(\kappa M) = \mu/D - f(\kappa M),$$

(4.5)

for some monotonic increasing function $f(\kappa M)$ having $f(0) = 0$, where κ represents the efficiency of use of metabolic energy. This equation has the equilibrium solutions (when $dD/dt = 0$)

$$D_{equlibG} = \mu/2f(\kappa M),$$

$$D_{equilibP} = \mu/f(\kappa M).$$

(4.6)

Here, for both Gaussian and Poisson processes, aging is represented by a decay in the efficiency of those chaperone processes, i.e., a slow decline in κ or increase in μ, that may involve idiosyncratic dynamics, ranging from punctuated phase transitions to autocatalytic runaway effects, since D, in equation (2.8), acts as a temperature analog for a system able to undergo punctuated spontaneous symmetry breaking phase transitions.

The reader is encouraged to explore the results for channels having rate distortion functions like equation (2.5),

$$R(D) = a \log[b/G_n(D)], 0 < D \leq G_n^{-1}(b); a, b, n > 0,$$

and G_n is a polynomial of order n that is zero only for $D = 0$ to ensure convexity in D.

4.2.2   A simple metabolic model

Again, the 'dual' treatment focuses on the rate distortion function $R(D)$, assuming that the probability density function for $R(D)$ at a given intensive index of embedding metabolic energy, M, can be described using an approach like equations (2.8) and (2.11):

$$Pr[R, \kappa M] = \frac{\exp[-R/Q(\kappa M)]}{\int_0^\infty \exp[-R/Q(\kappa M)]dR}$$

(4.7)

where $Q(\kappa M)$ is a monotonic increasing function with $Q(0) = 0$ that represents the synergism between the intensity and physiological availability of the embedding free energy. At a fixed value of κM, again taking a life course timeframe as opposed to a folding timeframe, the mean of R is

$$< R >= \int_0^\infty RPr[R, \kappa M]dR = Q(\kappa M).$$

(4.8)

A decline in κ can, again, trigger complicated phase change dynamics for this system, as R itself, according to equation (2.11), can act as a temperature analog in a symmetry breaking argument, causing sudden, punctuated, changes in the underlying protein folding mechanisms.

Solving for M in terms of $< R >$,

$$M = Q^{-1}(< R >)/\kappa.$$

(4.9)

Note that this argument leads to something much like the circumstance of figure 2.1. For small κ, or a particular form of Q, significant increase in $< R >$ can require explosive levels of available metabolic energy. For example, if

$$< R >= Q(\kappa M) = \sqrt[n]{\kappa M},$$

then

$$M = <R>^n / \kappa.$$

Similarly, for

$$<R> = \log[\kappa M + 1],$$

we have

$$M = \frac{\exp[<R>] - 1}{\kappa}.$$

Indeed, by the convexity of $R(D)$, assuming a singularity at $D = 0$, this qualitative result holds for all possible means of biological information transmission, generalizing the results of Lestas et al. (2010).

Note that taking a proper nonequilibrium Onsager relation, e.g., for the Gaussian channel,

$$dD/dt = -\mu dS_R/dD - [\mu/(2\sigma^2)] \exp[2Q(\kappa M)],$$

(4.10)

gives

$$R_{eq} \equiv R(D_{eq}) = Q(\kappa M),$$

(4.11)

and similarly for the Poisson channel, so that the two approaches are indeed dual.

Again, more general channels can be described using rate distortion functions of different forms, e.g., $R(D) = a \log[b/G_n(D)]$, where $a > 0$, G_n is a polynomial of order n in D, by convexity, singular at most at 0, and $b > 0$ is the noise parameter.

4.3   Intrinsically Disordered Proteins II

Equations (2.8) and (2.9) expressed the energetics of the protein folding funnel in terms of a distortion measure, D, generating a free energy Morse function

$$F_R = -\alpha D \log[\sum_i \exp[-\mathcal{H}_i/\alpha D]].$$

where $\mathcal{H}_i$ is a conformational free energy and α a scaling parameter.

A spontaneous symmetry breaking argument generates topological transitions in folded protein structure as D declines. As the channel capacity connecting the codon machine with amino acid/protein folding machines increases, the system is driven to a final low free energy 'frozen' conformation.

Taking the perspective of the previous sections, we calculate a metabolically-driven average $< D >$ for the Gaussian and Poisson channels using the relation

$$< D >= \frac{\int_0^{max} D \exp[-R(D)/Q(\kappa M)](dR/dD)dD}{\int_0^{max} \exp[-R(D)/Q(\kappa M)](dR/dD)dD}.$$

(4.12)

For the Gaussian and Poisson channels respectively,

$$< D_G >= \frac{\sigma^2}{2Q(\kappa M) + 1},$$

$$< D_P >= \frac{1/\lambda}{Q(\kappa M) + 1},$$

(4.13)

where, again, $Q(\kappa M)$ is monotonic increasing with $Q(0) = 0$.

Crudely, replacing D by $< D >$ in F_R gives explicit dependence on σ^2 or $1/\lambda$:

$$F_R \approx -\alpha < D_X > \log[\sum_i \exp[-\mathcal{H}_i/\alpha < D_X >]],$$

(4.14)

where $X = G, P$.

It is plausible, then, that the snake-like self-lubrication of the IDP decreasing both σ^2 and $1/\lambda$ may serve to lower the metabolic cost of certain complicated protein folding, signaling, regulatory, or related processes, thus markedly increasing fitness.

For completeness, it is worth noting that, regarding the dual relation in terms of the Rate Distortion Function, we would write

$$F_G \approx -\beta < R > \log[\sum_i \exp[-H_{G_i}/\beta < R >]],$$

(4.15)

where, from equation (4.8), $< R >= Q(\kappa M)$, so that

$$F_G \approx -\beta Q(\kappa M) \log[\sum_i \exp[-H_{Gi}/\beta Q(\kappa M)]].$$

(4.16)

Then the richness of possible structure is directly driven by the increasing 'temperature' of available metabolic free energy, providing a different portrait.

Recall, however, that, for the Poisson and Gaussian channels, $R \propto \log[(1/\lambda)/D], \log[\sigma^2/D]$ represents the minimum channel capacity needed to achieve an average distortion D. Thus self-lubrication by intrinsically disordered proteins that reduces σ^2 or $1/\lambda$ decreases the metabolic energy needed for a given symmetry richness.

4.4   A little Morse Theory

It seems possible to explore the underlying theory using standard topological arguments. Equations (2.8) and (2.14) have the same free energy form, although dual in a fundamental sense:

$$F = -\alpha D \log[\sum_i [\exp[-\frac{\mathcal{H}_i}{\alpha D}]],$$

$$F_G = -\beta R \log[\sum_i [\exp[\frac{-H_{Gi}}{\beta R}]].$$

D and R are, respectively, average distortion and the Rate Distortion Function, while $\mathcal{H}$ is a direct energy measure, and H_G an information source uncertainty.

The essential idea is that these are Morse functions, and that, in the sense of Pettini (2007), they index a topological hypothesis: Singularities in these free-energy analogs can be associated with a change in the topology of the underlying manifolds – sudden changes in the conformation of a folding protein.

Morse theory examines relations between analytic behavior of a function – the location and character of its critical points – and the underlying topology of the manifold on which the function is defined. Here we follow closely Pettini (2007).

The essential idea of Morse theory is to examine an n-dimensional manifold M as decomposed into level sets of some function $f : M \to \mathbf{R}$ where $\mathbf{R}$ is the set of real numbers.

The a-level set of f is defined as

$$f^{-1}(a) = \{x \in M : f(x) = a\},$$

the set of all points in M with $f(x) = a$. If M is compact, then the whole manifold can be decomposed into such slices in a canonical fashion between two limits, defined by the minimum and maximum of f on M. Let the part of M below a be defined as

$$M_a = f^{-1}(-\infty, a] = \{x \in M : f(x) \leq a\}.$$

These sets describe the whole manifold as a varies between the minimum and maximum of f.

Morse functions are defined as a particular set of smooth functions $f : M \to \mathbf{R}$ as follows. Suppose a function f has a critical point x_c, so that the derivative $df(x_c) = 0$, with critical value $f(x_c)$. Then f is a Morse function if its critical points are nondegenerate in the sense that the Hessian matrix of second derivatives at x_c, whose elements, in terms of local coordinates are

$$\mathcal{H}_{i,j} = \partial^2 f / \partial x^i \partial x^j,$$

has rank n, which means that it has only nonzero eigenvalues, so that there are no lines or surfaces of critical points and, ultimately, critical points are isolated.

The index of the critical point is the number of negative eigenvalues of $\mathcal{H}$ at x_c.

A level set $f^{-1}(a)$ of f is called a critical level if a is a critical value of f, that is, if there is at least one critical point $x_c \in f^{-1}(a)$.

Again following Pettini (2007), the essential results of Morse theory are:

[1] If an interval $[a, b]$ contains no critical values of f, then the topology of $f^{-1}[a, v]$ does not change for any $v \in (a, b]$. Importantly, the result is valid even if f is not a Morse function, but only a smooth function.

[2] If the interval $[a, b]$ contains critical values, the topology of $f^{-1}[a, v]$ changes in a manner determined by the properties of the matrix H at the critical points.

[3] If $f : M \to \mathbf{R}$ is a Morse function, the set of all the critical points of f is a discrete subset of M, i.e., critical points are isolated. This is Sard's Theorem.

[4] If $f : M \to \mathbf{R}$ is a Morse function, with M compact, then on a finite interval $[a, b] \subset \mathbf{R}$, there is only a finite number of critical points p of f such that $f(p) \in [a, b]$. The set of critical values of f is a discrete set of $\mathbf{R}$.

[5] For any differentiable manifold M, the set of Morse functions on M is an open dense set in the set of real functions of M of differentiability class r for $0 \le r \le \infty$.

[6] Some topological invariants of M, that is, quantities that are the same for all the manifolds that have the same topology as M, can be estimated and sometimes computed exactly once all the critical points of f are known: Let the Morse numbers $\mu_i (i = 0, ..., m)$ of a function f on M be the number of critical points of f of index i, (the number of negative eigenvalues of H). The Euler characteristic of the complicated manifold M can be expressed as the alternating sum of the Morse numbers of any Morse function on M,

$$\chi = \sum_{i=1}^{m} (-1)^i \mu_i.$$

The Euler characteristic reduces, in the case of a simple polyhedron, to

$$\chi = V - E + F$$

where V, E, and F are the numbers of vertices, edges, and faces in the polyhedron.

[7] Another important theorem states that, if the interval $[a, b]$ contains a critical value of f with a single critical point x_c, then the topology of the set M_b defined above differs from that of M_a in a way which is determined by the index, i, of the critical point. Then M_b is homeomorphic to the manifold obtained from attaching to M_a an i-handle, i.e., the direct product of an i-disk and an $(m - i)$-disk.

Again, see Pettini (2007) or Matsumoto (2002) for details.

An example is shown in figure 4.1, on the torus, where singularities in $f(p)$ occur at points $p_0...p_3$. p would, in our case, represent some function of D or R, acting as a temperature-anolog.

Note that point [6] produces Tlusty's fundamental quantity, $V - E + F$ of equation (1.2), in terms of Morse function singularities, opening the way for considerable theoretical development.

4.5   More Morse Functions

Although the Gibbs distribution seems to work well in describing protein folding dynamics in equation (2.18), it does not really seem altogether appropriate for a system whose dynamics evolve in an open manner. However, as we have argued, the regularities imposed by the asymptotic limit theorems of information theory permit study of 'nonequilibrium equilibria' in a standard way. Here we extend the treatment, adopting a perspective from network information theory (e.g., Cover and Thomas, 1991; El Gamal and Kim, 2010). That theory is, however, much a work in progress, with many unsolved difficulties. As El Gamal and Kim note, the simplistic model of a network consisting of separate links and naive forwarding nodes does not capture many important aspects of real world networked systems that involve multiple sources with various messaging requirements, redundancies, time and space correlations, and time variations. As they note, the goal in many information systems is not merely to communicate source information, but to make a decision or coordinate an action – in our context, cognitive process. Indeed, the first paper on network information theory was by Claude Shannon himself, who did not solve the question of optimal rates, a matter that remains open (Shannon, 1961), along with many others.

We suppose that the 'temperature' function – Q, R, D, or whatever – *is itself associated with an information source, Z*. This source represents an identifiable subset of the 'environmental' dynamics, in a large sense, and provides an embedding context for dynamic process. It defines *jointly typical* paths (Cover and Thomas, 2006) for an associated set of organismal information sources.

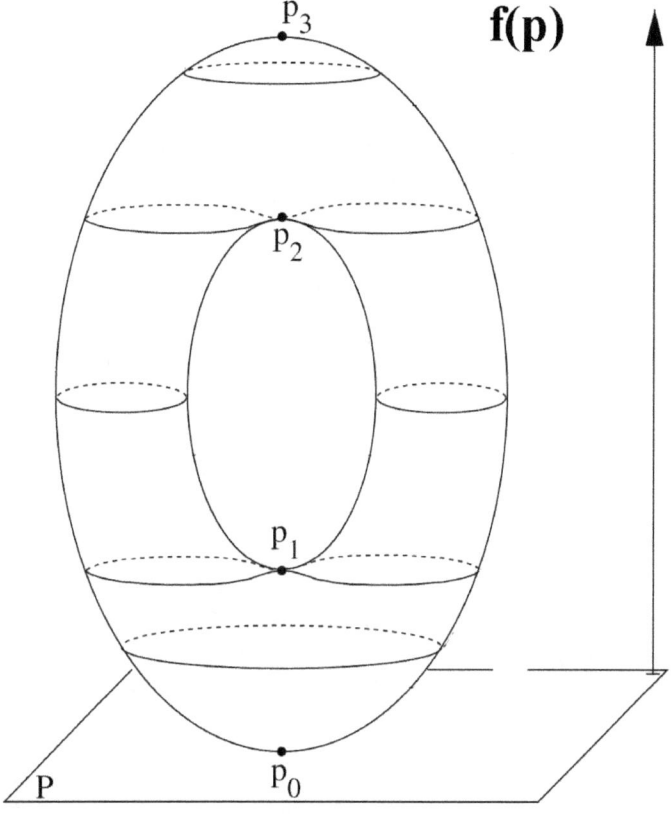

Figure 4.1: Adapted from Pettini (2007). A Morse function, $f(p)$, on the torus. Singularities in $f(p)$ occur at the indicated points, as p increases, representing the topological structure of the torus. A sphere would have only two singularities, top and bottom.

Given three interacting information sources, Y_1, Y_2, Z, the splitting criterion for tripartite jointly typical sequences, taking Z as an external context, is (Cover and Thomas, 1991)

$$I(Y_1; Y_2|Z) = H(Z) + H(Y_1|Z) + H(Y_2|Z) - H(Y_1, Y_2, Z),$$

(4.17)

where $H(...|...)$ and $H(..., ..., ...)$ represent conditional and joint uncertainties (Ash, 1990; Khinchin, 1957; Cover and Thomas, 1991).

This presumably generalizes to something like

$$I(Y_1; ...; Y_n|Z) = H(Z) + \sum_{j=1}^{n} H(Y_j|Z) - H(Y_1, ..., Y_n, Z).$$

(4.18)

More complicated multivariate typical sequences receive much the same treatment (El Gamel and Kim, 2010, p.2-26). Given a basic set of information sources $(X_1, ..., X_k)$ that one partitions into two ordered sets $X(\mathcal{J})$ and $X(\mathcal{J}')$, then the splitting criterion becomes $H(X(\mathcal{J})|X(\mathcal{J}'))$. Generalization to three or more ordered sets is straightforward.

Then the joint splitting criterion – I, H above – however it may be expressed as a composite of the underlying information sources and their interactions, satisfies a relation closely analogous to the first one in equation (2.7), where $N(n)$ is the number of high probability jointly typical paths of length n. Here the joint splitting criterion *is given as a functional composition of the underlying information sources and their interactions.*

The key point is that I in equation (4.18) and its generalizations in terms of $H(X(\mathcal{J})|X(\mathcal{J}'))$ etc. can themselves be considered as Morse Functions that can be parameterized in terms of the monotonic expression involving the temperature analogs Q, R, etc. The natural association of equivalence classes of states and trajectories with groupoid symmetries then suggests that Landau's spontaneous symmetry breaking arguments, extended to groupoids, will again apply. This can produce richer and more 'symmetric' biological processes and structures as

Q, R, etc. increase, leading to a sequence of 'eukaryotic-like' biological transitions.

I in equation (4.18) and the other splitting criteria analogous to it, however, have, in a sense, a more 'natural' interpretation than a Gibbs-based F_G: The inference is that choice of a proper Morse Function may depend strongly on biological context, with a simple Gibbs distribution sufficient for strongly 'physics-bound' processes such as protein folding, while more complex splitting criteria are to be associated with more complex biological phenomena.

Chapter 5

Alzheimer's, aging, and stress

A classic paper by Blackburn's group (Epel et al., 2004) detected accelerated telomere shortening in response to life stress, a powerful indication of early aging. Numerous studies demonstrate links between chronic stress and indices of poor health, including risk factors for cardiovascular disease and poorer immune function. More directly, Epel et al. found that psychosocial stress, both perceived stress and chronicity of stress, is significantly associated with higher oxidative stress, lower telomerase activity, and shorter telomere length – all known determinants of cell senescence and longevity – in peripheral blood mononuclear cells from healthy premenopausal women. Those with the highest levels of perceived stress had telomeres shorter on average by the equivalent of at least one decade of additional aging compared to low stress women. Thus, at the cellular level, stress may promote earlier onset of age-related diseases.

Somewhat more recent work relates physical and psychosocial stress to the etiology of Alzheimer's disease. Dong and Csernansky (2009) find growing evidence that the effects of such stressors on the hypothalamic-pituitary-adrenal (HPA) axis may be implicated in the onset and progression of Alzheimer's disease (AD). For instance, increases in plasma cortisol levels have been reported in patients diagnosed with AD. Moreover, they argue, a correlation has been reported between increases in 24 h cortisol levels and the severity of cognitive deficits in AD patients and between elevations in basal cortisol levels and the frequency of dexamethasone non-suppression in AD. Furthermore, postmortem cerebrospinal fluid (CSF) cortisol levels in AD patients were 83 % higher than in controls. Finally, changes in HPA axis in AD do not appear to be secondary to depression.

But, they conclude, while there is growing evidence that stress has an important impact on the pathogenesis of AD, the cellular mechanisms that might link stress and AD are not well understood.

A recent meta-analysis by the Alzheimer's Association (AA, 2006), found that, among studies of African-Americans, age-specific AD rates were from 14 to 100 percent higher than for matched White populations. Gurland et al. (1999) and Tang et al. (2001) observed the highest rates of AD among African-Americans and Hispanic elderly, when compared with non-Hispanic Whites in a relatively stable Upper Manhattan neighborhood.

Our hypothesis, of course, is that cortisol and other indices of HPA axis activity imply chemical stress on the protein regulatory apparatus, and serve as agents of premature aging. But HPA axis activity, as the work of Epel et al. (2004) indicates, is not a simple physiological response in humans. Rather, what constitutes stress is psychosocially and culturally interpreted.

5.1   The simplest HPA axis model

Stress, as we envision it, is not a random sequence of perturbations, and is not independent of its perception. Rather, it involves a highly correlated, grammatical, syntactical process by which an embedding psychosocial environment communicates with an individual, particularly with that individual's HPA axis, in the context of social hierarchy. We view the stress experienced by an individual as an adiabatically piecewise stationary ergodic (APSE) information source, interacting with a similar dual information source defined by HPA axis cognition, in the sense of Chapter 3.

The ergodic nature of the 'language' of stress – it's essential characteristic as an information source – is a generalization of the law of large numbers, so that long-time averages can be well approximated by cross-sectional expectations. Languages do not have simple autocorrelation patterns, in distinct contrast with the usual assumption of random perturbations by white noise in the standard formulation of stochastic differential equations.

Let us suppose we cannot measure stress, but can determine the concentrations of HPA axis hormones and other biochemicals according to some natural time frame, that we will characterize as the inherent period of the system. Suppose, in the absence of extraordinary meaningful psychosocial stress, we measure a series of n concentrations at time t which we represent as an n-dimensional vector X_t. Suppose we conduct a number of experiments, and create a regression model so that we can, in the absence of perturbation, write, to first order, the concentration of biomarkers at time $t + 1$ in terms of that at time t using a matrix equation of the form

$$X_{t+1} \approx < \mathbf{U} > X_t + b_0,$$

(5.1)

where $< \mathbf{U} >$ is the matrix of regression coefficients and b_0 a vector of constant terms.

We then suppose that, in the presence of a perturbation by structured stress

$$X_{t+1} = (< \mathbf{U} > +\delta \mathbf{U}_{t+1})X_t + b_0$$

$$\equiv < \mathbf{U} > X_t + \epsilon_{t+1},$$

(5.2)

where we have absorbed both b_0 and $\delta \mathbf{U}_{t+1} X_t$ into a vector ϵ_{t+1} of error terms that are not necessarily small in this formulation. In addition it is important to realize that this is not a population process whose continuous analog is exponential growth. Rather what we examine is more akin to the passage of a signal – structured psychosocial stress – through a distorting physiological filter.

If the matrix of regression coefficients $< \mathbf{U} >$ is sufficiently regular, we can (Jordan block) diagonalize it using the matrix of its column eigenvectors $\mathbf{Q}$, writing

$$\mathbf{Q}X_{t+1} = (\mathbf{Q} < \mathbf{U} > \mathbf{Q}^{-1})\mathbf{Q}X_t + \mathbf{Q}\epsilon_{t+1},$$

(5.3)

or equivalently as

$$Y_{t+1} =< \mathbf{J} > Y_t + W_{t+1},$$

(5.4)

where $Y_t \equiv \mathbf{Q}X_t, W_{t+1} \equiv \mathbf{Q}\epsilon_{t+1}$, and $< \mathbf{J} >\equiv \mathbf{Q} < \mathbf{U} > \mathbf{Q}^{-1}$ is a (block) diagonal matrix in terms of the eigenvalues of $< \mathbf{U} >$.

Thus the (rate distorted) writing of structured stress on the HPA axis through $\delta \mathbf{U}_{t+1}$ is reexpressed in terms of the vector W_{t+1}.

The sequence of W_{t+1} is the rate-distorted image of the information source defined by the system of external structured psychosocial stress. This formulation permits estimation of the long-term steady-state effects of that image on the HPA axis. The basic idea is to recognize that because everything is APSE, we can either time or ensemble average both sides of equation (5.4), so that the one-period offset is absorbed in the averaging, giving an 'equilibrium' relation

$$< Y >=< \mathbf{J} >< Y > + < W >$$

or

$$< Y >= (\mathbf{I} - < \mathbf{J} >)^{-1} < W >,$$

(5.5)

where $\mathbf{I}$ is the $n \times n$ identity matrix.

Now we reverse the argument: Suppose that Y_k is chosen to be some fixed eigenvector of $< \mathbf{U} >$. Using the diagonalization of $< \mathbf{J} >$ in terms of its eigenvalues, we obtain the average excitation of the HPA axis in terms of some eigentransformed pattern of exciting perturbations as

$$< Y_k >= \frac{1}{1 - < \lambda_k >} < W_k >$$

(5.6)

where $< \lambda_k >$ is the eigenvalue of $< Y_k >$, and $< W_k >$ is some appropriately transformed set of ongoing perturbations by structured psychosocial stress.

The essence of this result is that *there will be a characteristic form of perturbation by structured psychosocial stress – the W_k – that will resonantly excite a particular eigenmode of the HPA axis.* Conversely, by tuning the eigenmodes of $< \mathbf{U} >$, the HPA axis can be trained to galvanized response in the presence of particular forms of perturbation.

This is because, if $< \mathbf{U} >$ has been appropriately determined from regression relations, then the λ_k will be a kind of multiple correlation coefficient (Wallace and Wallace, 2000), so that particular eigenpatterns of perturbation will have greatly amplified impact on the behavior of the HPA axis. If $\lambda = 0$ then perturbation has no more effect than its own magnitude. If, however, $\lambda \to 1$, then the written image of a perturbing psychosocial stressor will have very great effect on the HPA axis. Following Ives (1995), we call a system with $\lambda \approx 0$ *resilient* since its response is no greater than the perturbation itself.

5.2   The generalized retina

Cohen (2000) argues for an 'immunological homunculus' as the immune system's perception of the body as a whole. The particular utility of such a concept, in his view, is that sensing perturbations in a bodily self-image can serve as an early warning sign of pending necessary inflammatory response – expressions of tumorigenesis, acute or chronic infection, parasitization, and the like. Thayer and Lane (2000) argue something analogous for emotional response as a quick internal index of larger patterns of threat or opportunity.

It seems obvious that the cognitive protein folding submodule we have invoked must also have a coherent internal self-image of the state of the protein folding process and its dynamics. This inferred picture we term a 'generalized retina' (GR). We shall use the responses of the GR to characterize responses to both dysfunction and to medical interventions, including drugs, used to treat that dysfunction. Clearly, illness and treatment may reflect one another in a hall of mirrors reminiscent of Jerne's idiotypic network proposed for the dynamics of the immune system.

Let us suppose we cannot measure either stress or cognitive submodule function directly, but can determine the concentrations of appropriate biomarkers, associated with the function of the cognitive protein folding system according to some natural time frame inherent to it. Suppose, in the absence of extraordinary meaningful psychosocial or other stress, we measure a series of n biomarker concentrations at time t, and represent them as an n-dimensional vector X_t. Suppose we conduct a number of experiments, and create a regression model so that we can, in the absence of perturbation, write, to first order, the markers at time $t+1$ in terms of that at time t using a matrix equation of the form

$$X_{t+1} \approx \mathbf{R}X_t,$$

(5.7)

where $\mathbf{R}$ is the matrix of regression coefficients, and we have normalized to a zero vector of constant terms.

Suppose we write a GR response to short-term perturbation as

$$X_{t+1} = (\mathbf{R}_0 + \delta\mathbf{R}_{t+1})X_t,$$

where $\delta\mathbf{R}$ represents variation of the generalized cognitive self-image about the basic state $\mathbf{R}_0$.

We impose a (Jordan block) diagonalization in terms of the matrix of (generally nonorthogonal) eigenvectors $\mathbf{Q}_0$ of some 'zero reference state' $\mathbf{R}_0$, obtaining, for an initial condition which is an eigenvector $Y_t \equiv Y_k$ of $\mathbf{R}_0$,

$$Y_{t+1} = (\mathbf{J}_0 + \delta\mathbf{J}_{t+1})Y_k = \lambda_k Y_k + \delta Y_{t+1} =$$

$$\lambda_k Y_k + \sum_{j=1}^{n} a_j Y_j,$$

(5.8)

where $\mathbf{J}_0$ is a (block) diagonal matrix as above, $\delta\mathbf{J}_{t+1} \equiv \mathbf{Q}_0\delta\mathbf{R}_{t+1}\mathbf{Q}_0^{-1}$, and δY_{t+1} *has been expanded in terms of a spectrum of the eigenvectors of* $\mathbf{R}_0$, with

$$|a_j| \ll |\lambda_k|, |a_{j+1}| \ll |a_j|.$$

(5.9)

The essential point is that, provided R_0 has been properly tuned, so that this condition is true, the first few terms in the spectrum of the plieotropic iteration of the eigenstate will contain almost all of the essential information about the perturbation, i.e., most of the variance. We envision this as similar to the detection of color in the optical retina, where three overlapping non-orthogonal eigenmodes of response suffice to characterize a vast array of color sensations. Here, if a concise spectral expansion is possible, a very small number of (typically nonorthogonal) generalized cognitive eigenmodes permit characterization of a vast range of external perturbations, and rate distortion constraints become very manageable indeed. Thus GR responses – the spectrum of excited eigenmodes of R_0, provided it is properly tuned – can be a very accurate and precise index of 'environmental' perturbation.

The choice of zero reference state R_0, the 'base state' from which perturbations are measured, is, we claim, a highly nontrivial task, necessitating a specialized apparatus.

This is no small matter. According to current theory, the adapted human mind functions through the action and interaction of distinct mental modules that evolved fairly rapidly to help address special problems of environmental and social selection pressure faced by our Pleistocene ancestors (Barkow et al., 1992). Here we have postulated the necessity of a cognitive module for protein folding regulation. As is well known in computer engineering, calculation by specialized submodules – numeric processor chips – can be a far more efficient means of solving particular well-defined classes of problems than direct computation by a generalized system. We suggest, then, that the generalized cognition needed for the regulation of protein folding has evolved specialized submodules to speed the address of certain commonly recurring challenges. Nunney (1999) has argued that, as a power law of cell count, specialized subsystems are increasingly required to recognize and redress tumorigenesis for different tissue types, mechanisms ranging from molecular error-correcting codes, to programmed cell death, and finally full-blown immune attack.

We argue that identification of the designated normal state of the GR – a cognitive self-image of the state of protein folding – is difficult, requiring a dedicated cognitive submodule within the overall generalized protein folding regulation cognitive machinery. This is essentially because, for the vast majority of information systems, unlike mechanical systems, there are no restoring springs whose low energy state automatically identifies equilibrium: relatively speaking, all states of any GR are 'high energy' states. That is, active comparison must be made of the state of the GR with some stored internal reference picture, and a decision made about whether to reset to zero, which is a cognitive process. We further speculate that the complexity of such a submodule may also follow something like Nunney's power law with physiological complexity, in some appropriate manner.

Failure of that cognitive submodule could result in identification of an excited state of the GR as normal, triggering patterns of systemic activation that could contribute to certain forms of protein folding disorder. This would result in a relatively small number of characteristic eigenforms of comorbidity, that would typically become more mixed with increasing disorder.

In sum, since such 'zero mode identification' (ZMI) is a (presumed) cognitive submodule of overall generalized protein folding cognition, it involves convoluting incoming 'sensory' with 'ongoing' internal memory data in choosing the zero state, i.e., defining $\mathbf{R_0}$. The dual information source defined by this cognitive process can then interact in a punctuated manner with 'external information sources' according to the Rate Distortion and related arguments above. From a RDT perspective, then, those external information sources literally write a distorted image of themselves onto the ZMI, often in a punctuated manner: (relatively) sudden onset of a developmental trajectory to particular sets of comorbid disorders.

Different systems of external signals – including but not limited to structured psychosocial stress – may write different characteristic images of themselves onto the ZMI cognitive submodule, i.e., trigger different patterns of comorbid protein disorder, e.g., type 2 diabetes and Alzheimer's disease.

A brief reformulation in abstract terms may be of interest. Recall that the essential characteristic of cognition in our formalism involves a function h that maps a (convolutional) path $x = a_0, a_1, ..., a_n, ...$ onto a member of one of two disjoint sets, B_0 or B_1. Thus respectively, either (1) $h(x) \in B_0$, implying no action taken, or (2), $h(x) \in B_1$, and some particular response is chosen from a large repertoire of possible responses. There is an obvious problem of defining these two disjoint sets, and that may well require some 'higher order cognitive module' to identify what constituted B_0, the set of 'normal' states. Again, this is because there is no low energy mode for information systems: virtually all states are more or less high energy states, and there is no way to identify a ground state using the physicist's favorite variational or other minimization arguments on energy.

Suppose that higher order cognitive module, which we now recognize as a kind of Zero Mode Identification, interacts with an embedding language of structured psychosocial stress (or other systemic perturbation, like a chemical exposure) and, instantiating a Rate Distortion image of that embedding stress, begins to include one or more members of the set B_1 into the set B_0. Recurrent 'hits' on that aberrant state would be experienced as episodes of highly structured comorbid protein folding dysfunction.

Empirical tests of this hypothesis, however, quickly lead again into real-world regression models involving the interrelations of measurable biomarkers, requiring formalism much like that used above. The GR can, then, be viewed as a generic heuristic device typifying such regression approaches.

The retina formalism is more appropriately characterized as a 'Rate Distortion Manifold', a local projection that, through overlap, has global structure, much like the tangent planes to a complicated geometric object. Glazebrook and Wallace (2009a, b) provide more detailed, indeed cutting-edge, mathematical treatment.

One inference from this model is that some (but not all) protein folding/aggregation disorders may be analogous to certain autoimmune diseases, in the sense that a pathological mode becomes misidentified as base-normal.

Another possible class may simply represent the overloading of regulatory systems.

5.3   Therapeutic failure

Socioculturally constructed and structured psychosocial stress, in this model having both (generalized) grammar and syntax, can be viewed as entraining the function of zero mode identification when the coupling with stress exceeds a threshold. More than one threshold appears likely, accounting in a sense for the typically staged nature of environmentally caused disorders. These should result in a synergistic – i.e., comorbidly excited – mixed affective, protein folding disorders, and represent the effect of stress on the linked decision processes of various cognitive functions, in particular through the identification of a false 'zero mode' of the GR. This is a collective, but highly systematic, 'tuning failure' that, in the Rate Distortion sense, represents a literal image of the structure of imposed psychosocial stress written upon the ability of the GR to characterize a normal condition of protein folding, causing a mixed excited state of chronically comorbid protein folding dysfunction.

In this model different eigenmodes Y_k of the GR regression model characterized by the matrix $\mathbf{R}_0$ can be taken to represent the 'shifting-of-gears' between different 'languages' defining the sets B_0 and B_1. That is, different eigenmodes of the GR would correspond to different required (and possibly mixed) characteristic systemic responses.

If there is a state (or set of states) Y_1 such that $\mathbf{R}_0 Y_1 = Y_1$, then the 'unitary kernel' Y_1 corresponds to the condition 'no response required', the set B_0, that is, normal function.

Suppose pathology becomes manifest, i.e.,

$$\mathbf{R}_0 \rightarrow \mathbf{R}_0 + \delta\mathbf{R} \equiv \hat{\mathbf{R}}_0,$$

so that some chronic dysfunctional state becomes the new 'unitary kernel', and

$$Y_1 \rightarrow \hat{Y}_1 \neq Y_1$$

$$\hat{\mathbf{R}}_0\hat{Y}_1 = \hat{Y}_1.$$

This could represent various forms of protein folding disorder.

Suppose we wish to induce a sequence of therapeutic counterperturbations $\delta\mathbf{T}_k$ according to the pattern

$$[\hat{\mathbf{R}}_0 + \delta\mathbf{T}_1]\hat{Y}_1 = Y^1,$$

$$\hat{\mathbf{R}}_1 \equiv \hat{\mathbf{R}}_0 + \delta\mathbf{T}_1,$$

$$[\hat{\mathbf{R}}_1 + \delta\mathbf{T}_2]Y^1 = Y^2$$

$$...$$

(5.10)

so that, in some sense,

$$Y^j \rightarrow Y_1.$$

(5.11)

That is, the protein folding system, as monitored by the GR, is driven to its original condition.

We may or may not have $\hat{\mathbf{R}}_0 \rightarrow \mathbf{R}_0$. That is, actual cure may not be possible, in which case palliation or control is the therapeutic aim.

The essential point is that the pathological state represented by $\hat{\mathbf{R}}_0$ and the sequence of therapeutic interventions $\delta \mathbf{T}_k, k = 1, 2, \ldots$ are interactive and reflective, depending on the regression of the set of vectors Y^j to the desired state Y_1, again, much in the same spirit as Jerne's immunological idiotypic hall of mirrors.

The therapeutic problem revolves around minimizing the difference between Y^k and Y_1 over the course of treatment: that difference represents the inextricable convolution of 'treatment failure' with 'adverse reactions' to the course of treatment itself, and 'failure of compliance' attributed through social construction by provider to patient, i.e., failure of the therapeutic alliance.

It should be obvious that the treatment sequence $\delta \mathbf{T}_k$ represents a cognitive path of interventions having, in turn, a dual information source in the sense we have previously invoked.

Treatment may, then, interact in the usual Rate Distortion manner with patterns of structured psychosocial stress that are, themselves, signals from an embedding information source. Thus treatment failure, adverse reactions, and patient noncompliance will, of necessity, embody a distorted image of structured psychosocial stress.

In sum, characteristic patterns of treatment failure, adverse reactions, and patient noncompliance reflecting collapse of the therapeutic alliance, will occur in virtually all therapeutic interventions according to the manner in which structured psychosocial stress is expressed as an image within the treatment process. This would most likely occur in a highly punctuated manner, depending in a quantitative way on the degree of coupling of the three-fold system of affected individual, patient/provider interaction, and treatment mode, with that stress.

Given that the principal environment of humans is defined by interaction with other humans and with socioeconomic institutions, these are likely to be very strong effects indeed, and we examine recent trajectories of psychosocial stress in the United States, and their effects on Alzheimer's mortality incidence at the state level, in the next chapter.

5.4 The Rate Distortion Manifold

The previous two sections revolved around the identification of a 'base operator' $\mathbf{R}_0$, representing a normal physiological condition that directs the dynamic behavior of an information system. Here we reformulate that idea in fundamental terms, using the 'tuning theorem' variant of the Shannon Coding Theorem.

5.4.1 The Coding Theorem

Messages from a source, seen as symbols x_j from some alphabet, each having probabilities P_j associated with a random variable X, are 'encoded' into the language of a 'transmission channel', a random variable Y with symbols y_k, having probabilities P_k, possibly with error. Someone receiving the symbol y_k then retranslates it (without error) into some x_k, which may or may not be the same as the x_j that was sent.

More formally, the message sent along the channel is characterized by a random variable X having the distribution

$$P(X = x_j) = P_j, j = 1, ..., M.$$

The channel through which the message is sent is characterized by a second random variable Y having the distribution

$$P(Y = y_k) = P_k, k = 1, ..., L.$$

Let the joint probability distribution of X and Y be defined as

$$P(X = x_j, Y = y_k) = P(x_j, y_k) = P_{j,k}$$

and the conditional probability of Y given X as

$$P(Y = y_k | X = x_j) = P(y_k | x_j).$$

Then the Shannon uncertainty of X and Y independently and the joint uncertainty of X and Y together are defined respectively as

$$H(X) = - \sum_{j=1}^{M} P_j \log(P_j)$$

$$H(Y) = - \sum_{k=1}^{L} P_k \log(P_k)$$

$$H(X,Y) = - \sum_{j=1}^{M} \sum_{k=1}^{L} P_{j,k} \log(P_{j,k}).$$

(5.12)

The *conditional uncertainty* of Y given X is defined as

$$H(Y|X) = -\sum_{j=1}^{M}\sum_{k=1}^{L} P_{j,k} \log[P(y_k|x_j)]$$

(5.13)

For any two stochastic variates X and Y, $H(Y) \geq H(Y|X)$, as knowledge of X generally gives some knowledge of Y. Equality occurs only in the case of stochastic independence.

Since $P(x_j, y_k) = P(x_j)P(y_k|x_j)$, we have

$$H(X|Y) = H(X,Y) - H(Y)$$

The information transmitted by translating the variable X into the channel transmission variable Y – possibly with error – and then re-translating without error the transmitted Y back into X is defined as

$$I(X|Y) \equiv H(X) - H(X|Y) = H(X) + H(Y) - H(X,Y)$$

(5.14)

See, for example, Ash (1990), Khinchin (1957) or Cover and Thomas (1991) for details. The essential point is that if there is no uncertainty in X given the channel Y, then there is no loss of information through transmission.

In general this will not be true, and herein lies the essence of the theory.

Given a fixed vocabulary for the transmitted variable X, and a fixed vocabulary and probability distribution for the channel Y, we may vary the probability distribution of X in such a way as to maximize the information sent. The capacity of the channel is defined as

$$C \equiv \max_{P(X)} I(X|Y)$$

(5.15)

subject to the subsidiary condition that $\sum P(X) = 1$.

The critical trick of the Shannon Coding Theorem for sending a message with arbitrarily small error along the channel Y at any rate $R < C$ is to encode it in longer and longer 'typical' sequences of the variable X; that is, those sequences whose distribution of symbols approximates the probability distribution $P(X)$ above which maximizes C.

If $S(n)$ is the number of such 'typical' sequences of length n, then

$$\log[S(n)] \approx nH(X)$$

where $H(X)$ is the uncertainty of the stochastic variable defined above. Some consideration shows that $S(n)$ is much less than the total number of possible messages of length n. Thus, as $n \to \infty$, only a vanishingly small fraction of all possible messages is meaningful in this sense. This observation, after some considerable development, is what allows the Coding Theorem to work so well. In sum, the prescription is to encode messages in typical sequences, which are sent at very nearly the capacity of the channel. As the encoded messages become longer and longer, their maximum possible rate of transmission without error approaches channel capacity as a limit. Again, Ash (1990), Khinchin (1957) and Cover and Thomas (1991) provide details.

5.4.2   The tuning theorem

Telephone lines, optical wave guides and the tenuous plasma through which a planetary probe transmits data to earth may all be viewed in traditional information-theoretic terms as a *noisy channel* around which we must structure a message so as to attain an optimal error-free transmission rate.

Telephone lines, wave guides and interplanetary plasmas are, relatively speaking, fixed on the timescale of most messages, as are most sociogeographic networks. Indeed, the capacity of a channel, according to equation (5.15), is defined by varying the probability distribution of the 'message' process X so as to maximize $I(X|Y)$.

Suppose there is some message X so critical that its probability distribution must remain fixed. The trick is to fix the distribution $P(x)$ but *modify the channel* – i.e. tune it – so as to maximize $I(X|Y)$. The *dual* channel capacity C^* can be defined as

$$C^* \equiv \max_{P(Y), P(Y|X)} I(X|Y)$$

(5.16)

But

$$C^* = \max_{P(Y),P(Y|X)} I(Y|X)$$

since

$$I(X|Y) = H(X) + H(Y) - H(X,Y) = I(Y|X).$$

Thus, in a purely formal mathematical sense, *the message transmits the channel*, and there will indeed be, according to the Coding Theorem, a channel distribution $P(Y)$ which maximizes C^*.

One may do better than this, however, by modifying the channel matrix $P(Y|X)$. Since

$$P(y_j) = \sum_{i=1}^{M} P(x_i)P(y_j|x_i),$$

$P(Y)$ is entirely defined by the channel matrix $P(Y|X)$ for fixed $P(X)$ and

$$C^* = \max_{P(Y),P(Y|X)} I(Y|X) = \max_{P(Y|X)} I(Y|X).$$

Calculating C^* requires maximizing the complicated expression

$$I(X|Y) = H(X) + H(Y) - H(X,Y)$$

which contains products of terms and their logs, subject to constraints that the sums of probabilities are 1 and each probability is itself between 0 and 1. Maximization is done by varying the channel matrix terms $P(y_j|x_i)$ within the constraints. This is a difficult problem in nonlinear optimization requiring Lagrange multiplier methods. However, for the special case $M = L$, C^* may be found by inspection: If $M = L$, then choose

$$P(y_j|x_i) = \delta_{j,i}$$

where $\delta_{i,j}$ is 1 if $i = j$ and 0 otherwise. For this special case

$$C^* \equiv H(X)$$

with $P(y_k) = P(x_k)$ for all k. *Information is thus transmitted without error when the channel becomes 'typical' with respect to the fixed message distribution $P(X)$.*

If $M < L$ matters reduce to this case, but for $L < M$ information must be lost, leading to Rate Distortion arguments explored more fully below.

Thus modifying the channel – "tuning $\mathbf{R}_0$ – may be a far more efficient means of ensuring transmission of an important message than encoding that message in a 'natural' language which maximizes the rate of transmission of information on a fixed channel.

We have examined the two limits in which either the distributions of $P(Y)$ or of $P(X)$ are kept fixed. The first provides the usual Shannon Coding Theorem, and the second, the tuning theorem variant. It seems likely, however, than for many important systems $P(X)$ and $P(Y)$ will 'interpenetrate,' to use Richard Levins' terminology. That is, $P(X)$ and $P(Y)$ will affect each other in characteristic ways, so that some form of mutual tuning may emerge as the most effective strategy.

The bottom line is that the physiological 'solutions' of the previous two sections, represented as $Y \to \mathbf{R}Y$, are simply the product of the information processing of a problem, and, by a very famous argument, information can never be gained simply by processing. Thus a 'problem' Y is transmitted as a message by an information processing channel, $\mathbf{R}$, a generalized computing device, and recoded as an 'answer' $\mathbf{R}Y$. By the Tuning Theorem argument, there will be a channel coding of $\mathbf{R}$ that, when properly tuned, is most efficiently transmitted by the 'problem' Y. In general, then, the most efficient coding of the transmission channel, that is, the best algorithm turning a problem into a solution, will necessarily be highly problem-specific. Thus there can be no best algorithm for all equivalence classes of problems, although there may well be an optimal algorithm for any given class.

Rate distortion, however, occurs when the problem is collapsed into a smaller, simplified, version and then solved. Then there must be a tradeoff between allowed average distortion and the rate of solution: the retina effect of a Rate Distortion Manifold. In a very fundamental sense – particularly for real time systems – rate distortion manifolds present a generalization of the converse of the famous no free lunch arguments (e.g., English, 1996). The neural corollary is known as inattentional blindness (Wallace, 2007).

See Glazebrook and Wallace (2009a, b) for a more extensive treatment of the Rate Distortion Manifold.

5.5   AD: some data

We suggest, following theory, that learning by the HPA axis is, in fact, the process of tuning response to perturbation. This is why one writes $< \mathbf{U} >$ instead of simply $\mathbf{U}$: The regression matrix is a tunable information system in the sense of the previous section. Different cultures can be expected to have different patterns of excitation – different tunings. That is, signals will excite HPA response in a culturally-specific manner. Thus the role of psychosocial stress in the etiology of HPA-related disorders, including Alzheimer's disease, can be culturally specific.

A recent study by Wilson et al. (2005) provides some data on cultural responses to stress in the etiology of Alzheimer's disease, for a population all over 65, with mean age 73.8, and half African-American:

> Persons without dementia residing in a biracial community completed a brief survey of proneness to psychological distress, and 1,064 were subsequently examined for incident Alzheimer's disease (AD) 3 to 6 years later. In analyses controlling for selected demographic and clinical variables, persons prone to distress were 2.4 times more likely to develop AD than persons not distress prone. This effect was substantially stronger in white persons compared to African Americans.

Figure 5.1, adapted from Wilson et al. (2005), compares rates of AD onset as a function of a survey measure of distress-proneness for White and African American subpopulations within their sample.

The results are consistent with a model of stress-induced aging of protein folding regulation that reflects cultural tuning of the HPA axis, in spite of the likely higher total burden of AD in the African-American population.

This result is, of course, at the individual scale of analysis and is critically dependent on the history of the particular 'biracial' community. Patterns of serial forced displacement and job loss affecting other African American communities, like those outlined in Wallace and Fullilove (2008), might well produce other patterns at larger scales of analysis – family, social network, community, and the like.

Some insight to the results of figure 5.1 can be gained from a recent spatial analysis of the relation between bodymass index, demoralization, and community stress by D. Wallace et al. (2003). That work studied a sample of several hundred pregnant women of Dominican (D) and African-American (AA) ethnicity in Upper Manhattan, from Inwood and Washington Heights, down through Central Harlem, attending a prenatal clinic. The women were overselected to exclude smokers and those with alcohol or drug dependencies, but included those willing to work closely with academic researchers. The neighborhoods were classified as Health Areas (HA), the small aggregates of census tracts by which the New York City Health Department reports data. These ranged from middle-class, primarily White HA's, through mixed, down into heavily segregated African-American sections. Where D and AA populations overlapped, there was no statistically significant difference

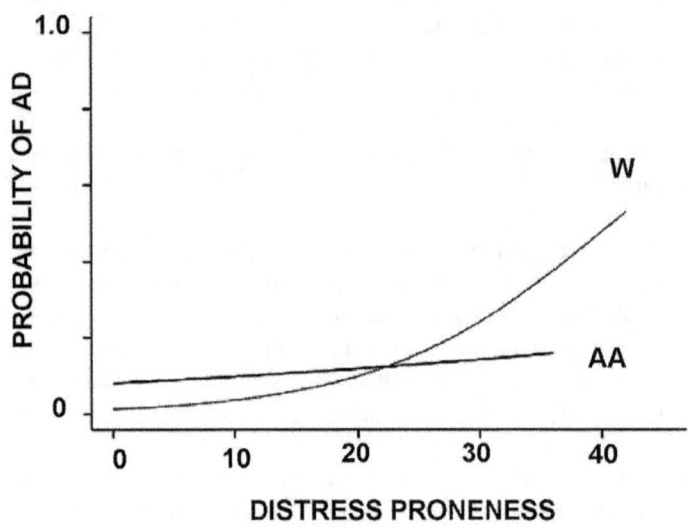

Figure 5.1: Adapted from Wilson et al. (2005). Probability of Alzheimer's disease as a function of Distress Proneness for White (W) and African-American (AA) subpopulations. Different responses to stress appear to represent fundamentally different cultural tunings of the HPA axis, at this individual level scale of study. Analysis at community scales, particularly in light of the serial forced displacements outlined in Wallace and Fullilove (2008), might show other patterns.

between the two ethnic groups in any demographic variate. Thus, for this particular sample, 'race is place', in the words of one researcher.

For the 27 Upper Manhattan HA's, an index of chronic community stress was constructed as the independent variable.

Incidence of selected socioeconomic and social-indicator health outcome factors were calculated for 1994-1996 by summing the case numbers for the three years and dividing by the area 1990 population. Because low-weight birth is well documented in the literature as associated with maternal stress, health area incidence of low-weight birth (1994-1996 number of births below 2500 g/10000 live births) was regressed against health area incidence of each chronic stressor. Those stressors significantly associated with low-weight birth incidence with an R^2 at or above 0.2 were included in the combined index. Thus, we included only stressors imposing a population level stress of potential importance.

These included:

Drug and Cirrhosis deaths per 100,000, unemployment rate, percent population on welfare, percent living in poverty, percent badly overcrowded housing, inverse median income, population change 1980-90, and percent foreign born.

Each stressor incidence of each HA was standardized by dividing it by the median incidence of the 27 health areas of the Upper Manhattan study zone. Each standardized incidence was weighted by multiplying it by the R^2 of its association with low-weight birth incidence. For each health area, the sum of all the standardized weighted incidences became the Index of Chronic Community Stress (ICCS). To check the validity of low-weight birth incidence as an indicator of stress, we also created an ICCS based on diabetes mortality incidence because type II diabetes (the vastly dominate type) has been correlated with community stress in the literature. Regression of the two indices of chronic stress yielded an R^2 of 0.99.

Dependent variables were bodymass index (BMI), a standard measure of obesity, and score on the standard CESD demoralization test. The HAs were aggregated into quintiles, based on their ICCS measure. Figure 5.2 shows the result.

The HAs having highest chronic stress index were located in Central Harlem, those with lowest stress in an affluent, primarily White, section of Inwood.

For this group of carefully chosen and extremely high function young women, we interpreted these results to indicate a marked shift in social network structure between low and high chronic stress HAs, with those living in Central Harlem, during the period of study, virtually housebound by high rates of street violence in the surrounding community. The lack of demoralization in that population was thus the product of a protective family/social network fragmentation, i.e., living behind

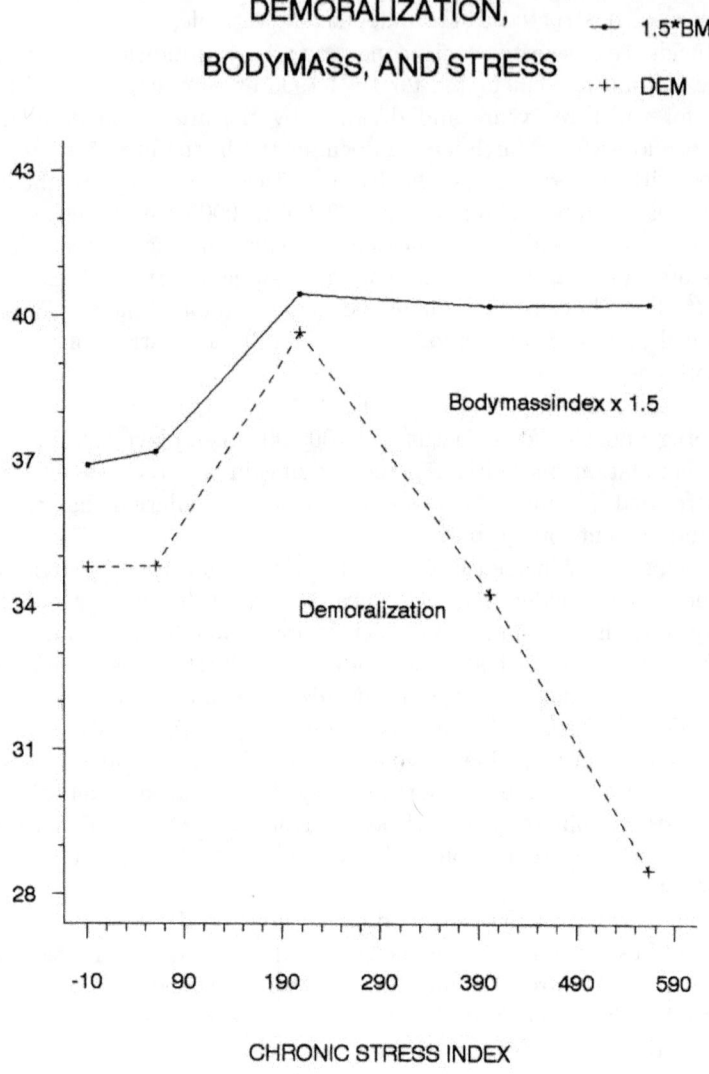

Figure 5.2: From D. Wallace et al., (2003). Average Demoralization Scores and average BMIs of the Chronic Community Stress Indices. Note the inverted 'U' shape with peak in the middle for the average demoralization scores. Note also the step function of the average BMIs with the high plateau for quintiles 3-5.

closed doors among friends and stable family members, isolated from the larger community and from individuals associated with substance abuse. The consequent high BMI, however, would, over the life course, constitute a serious burden to protein folding regulatory systems, quite possibly resulting in higher rates of Alzheimer's and related protein folding disorders during old age. The essential point being that whatever these disease rates, they would, for cultural reasons, not be associated with high CESD test scores, or with other formal measures of demoralization.

For the record, we interpreted the peak in demoralization score to represent the particular stress of 'living on the cliff' between White and Black zones, for this sample.

Thus, according to these results, tuning of the HPA axis can be induced by both environmental and cultural structures.

Chapter 6

The American catastrophe

6.1 Trajectories of stress

We know that both diabetes and hypertension are associated with increased risk for Alzheimer's disease (e.g., Sims-Robinson et al., 2010; Kivipelto et al., 2001). The 'obesity epidemic' in the US is related to rising rates of both disorders. Here we parse those increases according to White and African-American ethnicity, in accordance with the discussion above. This is not a pleasant task: Two powerful and intertwining phenomena of socioeconomic disintegration – deurbanization in the 1970's, and deindustrialization, particularly since 1980 – have combined to profoundly damage many US communities, dispersing historic accumulations of economic, political, and social capital. These losses have had manifold and persisting impacts on both institutions and individuals (Pappas, 1989; Ullmann, 1988; Wallace and Wallace, 1998). Elsewhere we examined the effect of these policy-driven phenomena on the hierarchical diffusion of AIDS in the US (Wallace et al., 1999). Here we extend that work to their association with obesity, in the context of the causal biological model for protein folding disorders given above, following the results of Wallace and Wallace (2005, 2010).

By 1980, not a single African-American urban community established before or during World War II remained intact. Many Hispanic urban neighborhoods established after the war suffered similar fates. Virtually all lost considerable housing, population, and economic and social capital either to programs of urban renewal in the 1950's or to policy-related contagious urban decay from the late 1960's through the late 1970's (Wallace and Wallace, 1998; Wallace and Fullilove, 2008).

Figure 6.1 gives an example, showing the percent change of occupied

PERCENT HOUSING LOST 1970–80

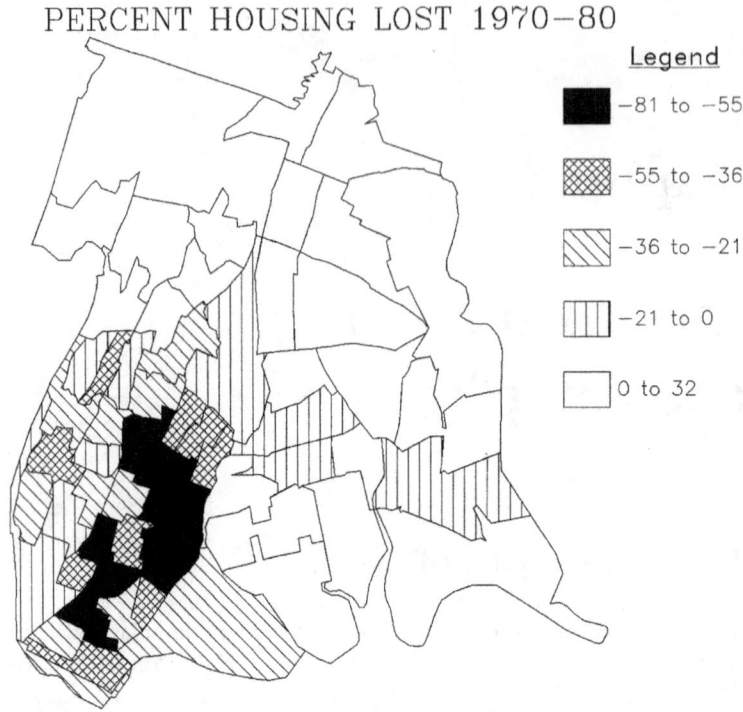

Figure 6.1: Percent change in occupied housing units, Bronx section of New York City, 1970-1980. Large areas lost over half their housing in this period, a degree of destruction unprecedented in an industrialized nation outside of wartime. Similar policy-driven disasters have afflicted most US urban minority communities since the end of World War II.

housing units in the Bronx section of New York City between 1970 and 1980 by Health Area, the aggregation of US Census Tracts by which morbidity and mortality are reported in the city. The South-Central section of the Bronx, by itself one of the largest urban concentrations in the Western world with about 1.4 million inhabitants, lost between 55 and 80 percent of housing units, most within a five year period. This is a level of damage unprecedented in an industrialized nation short of civil or international war, and indeed can be construed as a kind of covert civil war (Duryea, 1978; Wallace and Wallace, 1998).

Figure 6.2, a composite index of number and seriousness of building fires from 1959 through 1990 (Wallace and Wallace, 1998; Wallace et al., 1999), illustrates the process of contagious urban decay in New York City producing that housing loss, affecting large sections of

Harlem in Manhattan, and a broad band across the African-American and Hispanic neighborhoods of Northern Brooklyn, from Williamsburg to Bushwick, Brownsville, and East New York. The sudden rise between 1967 and 1968 was stemmed through 1972 by the opening of 20 new fire companies in high fire incidence, minority neighborhoods of the city. Beginning in late 1972, however, some 50 firefighting units were closed and many others destaffed as part of a 'planned shrinkage' program that continued the ethnic cleansing policies of 1950's urban renewal without benefit of either constitutional niceties or new housing construction to shelter the displaced population (Wallace and Wallace, 1998; Wallace and Fullilove, 2008).

Similar maps and graphs could be drawn for devastated sections of Detroit, Chicago, Los Angeles, Philadelphia, Baltimore, Cleveland, Pittsburgh, Newark, and a plethora of smaller US urban centers, each with its own individual story of active public policy and passive 'benign neglect'.

Figure 6.1 represents the Bronx part of the spatial distribution of the time integral of figure 6.2.

Figure 6.3, using data taken from the US Census, shows the counties of the Northeastern US losing more than 1000 manufacturing jobs between 1972 and 1987, the famous rust belt. It is, in its way, an exact parallel to figure 6.1 in that unionized manufacturing jobs lost remained lost, and their associated social capital and political influence were dispersed. As Pappas (1989) describes, the effects were profound and permanent:

> By 1982 mass unemployment had reemerged as a major social issue [in the USA]. Unemployment rose to its highest level since before World War II, and an estimated 12 million people were out of work – 10.8 percent of the labor force in the nation. It was not, however, a really new phenomenon. After 1968 a pattern was established in which each recession was followed by higher levels of unemployment during recovery. During the depth of the 1975 recession, national unemployment rose to 9.2 percent. In 1983, when a recovery was proclaimed, unemployment remained at 9.5 percent annually.
>
> Certain sectors of the work force have been more heavily affected than others. There was a 16.9 percent jobless rate among blue-collar workers in April, 1982... Unemployment and underemployment have become major problems for the working class. While monthly unemployment figures rise and fall, these underlying problems have persisted over a long period. Mild recoveries merely distract out attention from them.

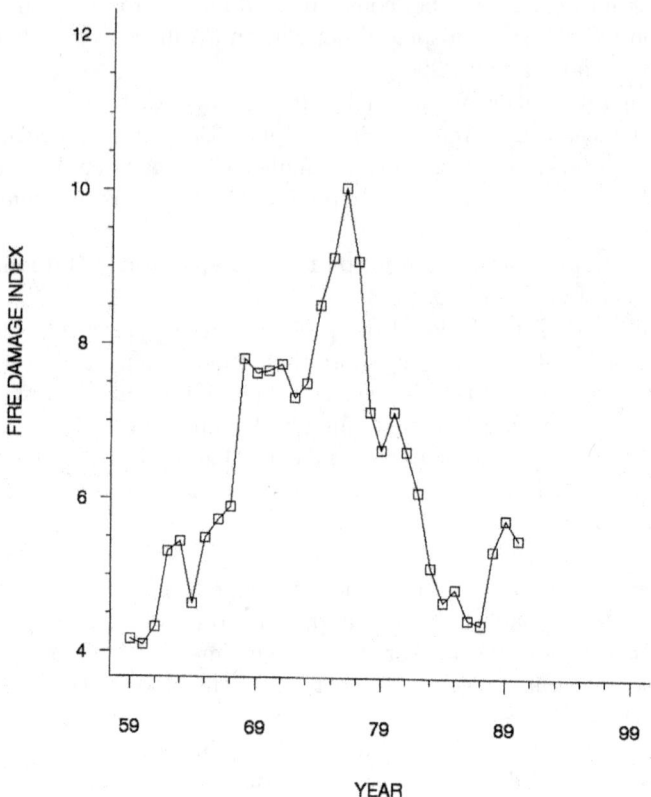

Figure 6.2: Annual fire damage index, New York City, 1959-1990, a composite of number and seriousness of structural fires, and an index of contagious urban decay. Some 20 new fire companies were added to high fire areas between 1969 and 1971, interrupting the process. Fifty firefighting units were closed in or permanently relocated from, high fire areas after November, 1972, allowing contagious urban decay to proceed to completion, producing the conditions of figure 6.1.

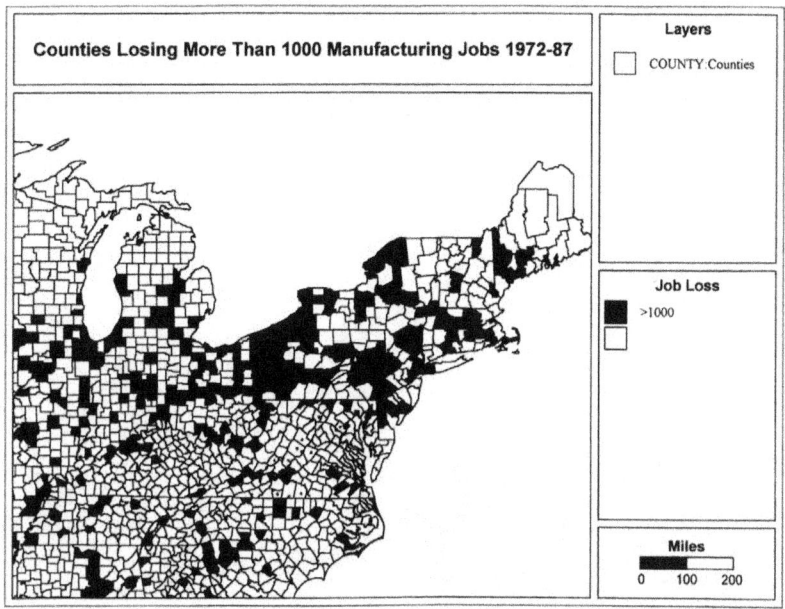

Figure 6.3: The Rust Belt: Counties of the Northeastern US which lost
1000 or more manufacturing jobs between 1972 and 1987.

Figure 6.4, using data from the US Bureau of Labor Statistics, shows
the total number of US manufacturing jobs from 1980 to 2001. We define our environmental index of the US national pattern of structured
stress to be represented by *the integral of manufacturing job loss after
1980*, i.e., the space between the observed curve and a horizontal line
drawn out from the 1980 number of jobs. This is not quite the same
as figure 6.3, that represents a simple net loss between two time periods. We believe that manufacturing job loss at one period continues to
have influence at subsequent periods as a consequence of permanently
dispersed social and political capital, at least over a 20 year span.

Other models, perhaps with different integral weighting functions,
are, of course, possible. We use simply

$$\mathcal{D}(T) = - \sum_{\tau=1980}^{\tau=T} [M(\tau) - M(1980)]$$

(6.1)

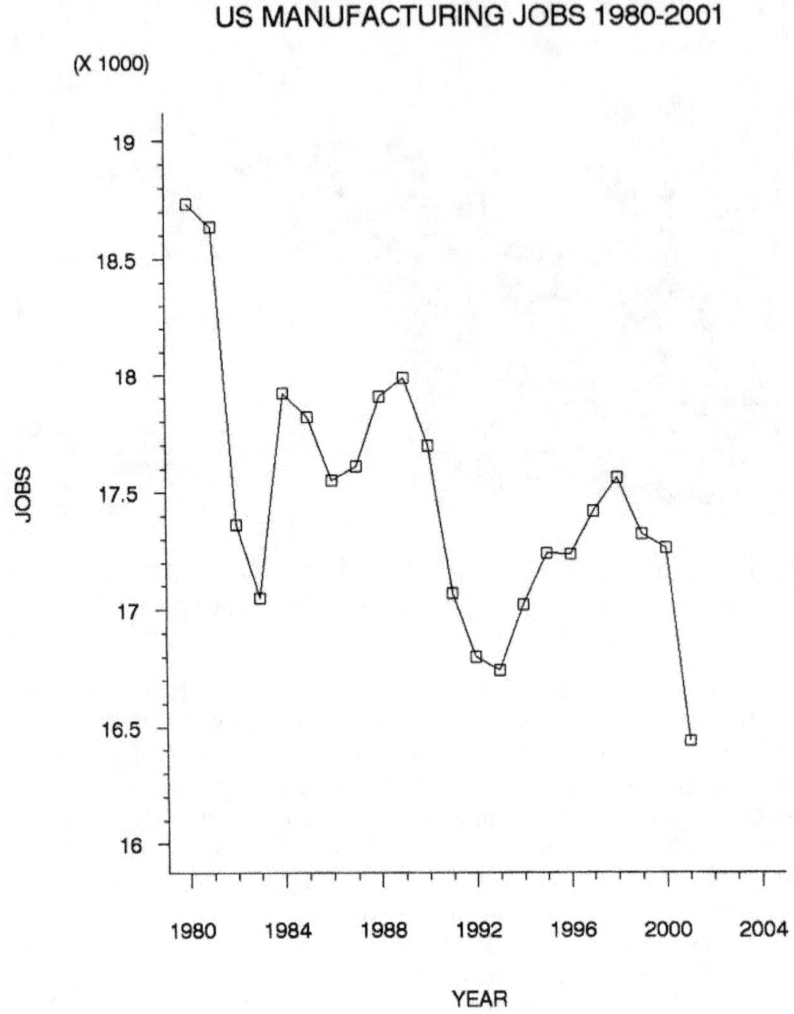

US MANUFACTURING JOBS 1980-2001

Figure 6.4: Annual number of manufacturing jobs in the US, 1980-2001. Our environmental index of social decay is the integrated loss after the 1980 peak, representing the permanent dispersal of economic, social, and political capital, part of the opportunity cost of a deindustrialization largely driven by the diversion of technical resources from civilian industry into the Cold War (e.g., Ullmann, 1988).

while a more elaborate treatment might involve something like

$$\mathcal{D}(T) = -\int_{\tau_0}^{T} f(T - \tau)M(\tau)d\tau$$

(6.2)

where $\mathcal{D}$ is the deficit, $M(\tau)$ is the number of manufacturing jobs at time τ, and $f(T - \tau)$ is a lagged weighting function.

Figure 6.5, using data from the Centers for Disease Control (2003), shows the percent of US adults characterized as obese according to the Behavioral Risk Factor Surveillance System between 1991 and 2001. This is given as a function of the integrated manufacturing jobs deficit from 1980, again, calculated as a simple negative sum of annual differences from 1980.

The association is quite good indeed, and the theory of the first sections suggests the relation is causal and not simply correlational: Loss of stable working class employment, loss of social and political capital, loss of union influence on working conditions and public policies, deurbanization intertwined with deindustrialization and their political outfalls, all constitute a massive threat expressing itself in population-level patterns of HPA axis-driven metabolism and metabolic syndrome.

Figure 6.6 extends the analysis to diabetes deaths in the US between 1980 and 1998. It shows the death rate per 100,000 as a function of the cumulative manufacturing jobs deficit from 1980 through 1998. Diabetes deaths are, after a lag, a good index of population obesity. Two systems are evident, before and after 1989, with a phase transition between them probably representing, in Holling's (1973) sense, a change in ecological domain roughly analogous to the sudden eutrophication of a lake progressively subjected to contaminated runoff. This would seem to reflect the delayed cumulative impacts of both deindustrialization and the deurbanization which became closely coupled with it. Further sudden, marked, upward transitions seem likely if socioeconomic and political reforms are not forthcoming. A simple linear correlation for the period gives an R^2 of 0.91, not inconsiderable.

It would be useful to compare annual county-level maps of diabetes, hypertension, Alzheimer's, and Parkinson's disease death rates with those of manufacturing job loss and deurbanization, but such a study would require considerable resources in order to conduct the necessary sophisticated analyses of cross-coupled, lagged, spatial and social diffusion.

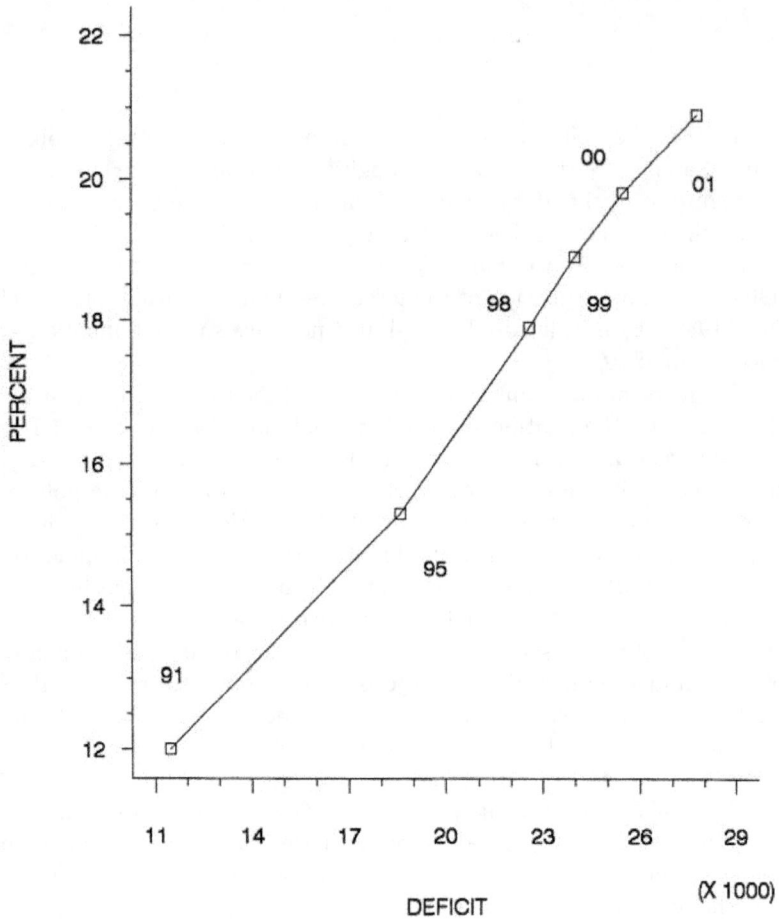

Figure 6.5: 1991-2001 relation between adult obesity in the US and the integrated loss of manufacturing jobs after 1980. We believe manufacturing job loss is an index of permanent decline in social, economic, and political capital that is perceived as, and indeed represents, a serious threat to the well-being of the US population.

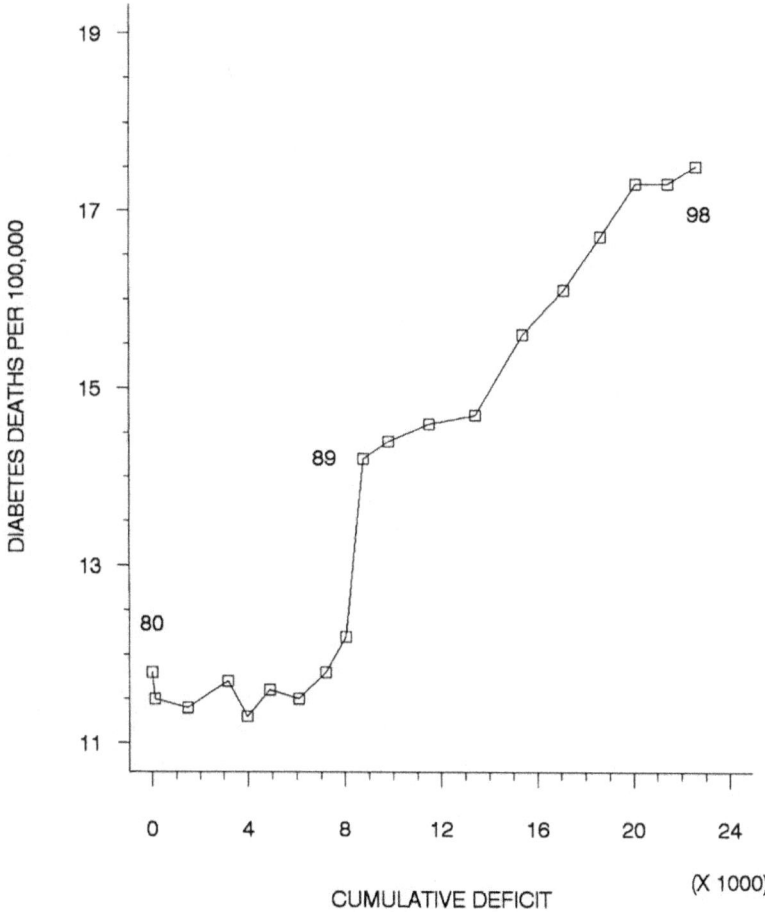

DIABETES DEATH RATE VS. CUMULATIVE
MANUFACTURING JOBS DEFICIT 1980-98

Figure 6.6: 1980-1998 relation between US diabetes death rate and integrated loss of manufacturing jobs after 1980. Two systems are evident: before and after 1989. We believe this sudden change represents a nonlinear transition between ecosystem domains which is much like the eutrophication of an increasingly contaminated water body (e.g., Holling, 1973). The simple correlation has $R^2 = 0.91$.

Figure 6.7 shows a regression of the Black vs. White diabetes death
rates (per 100,000) in the US for the period 1979-1997. It is striking
that, while the rate of increase for African-Americans was more than
50 % higher than for whites, both subpopulations were closely linked
together in a relentless progression: The R^2 of the regression was 0.99.
Similarly, figure 6.8 shows Black vs. White hypertension death rates
(per 100,000) over the same time span. Again, while African-Americans
suffered proportionally more than Whites, the two groups were closely
linked in a remarkable joint increase, $R^2 = 0.85$.

Diabetes and hypertension are, of course, both closely related to
obesity, and, as described above, to the onset of Alzheimer's disease.
The increased slopes of the regression lines of figures 6.7 and 6.8 indicate
that death rates of hypertension and diabetes in African-American populations are, respectively, 26 to 53 percent higher than for Whites over
the period 1979-1997, while both increase relentlessly during that period. This pattern suggests that rates of Alzheimer's disease in African-
Americans will be similarly, and necessarily, elevated (AA, 2006).

6.2   A pilot study

Two obvious national indices of the catastrophe implied by figures
6.3 and 6.4 are percent unemployed and the percent of the workforce in a union, and we construct an elementary model of state-level
Alzheimer's disease deaths based on them. The dependent variate is
the 'young elderly' annual average death rate (ICD-10 G30 classification) per 100,000 for the age range 65-74 in the period 1999-2006. Data
are taken from the CDC Wonder web site. The independent variates
are the percent unemployed in 2003, and percent of the workforce union
members, 2004, available from the US Department of Labor Statistics.

The 50-state regression model, having F=7.98, P=0.0010, adjusted
R^2=22.2 percent, is given by

Parameter	Estimate	SE	t	P
CONSTANT	18.76	3.86	4.86	0.0000
unemp03	1.407	0.694	2.03	0.048
unionpc04	-0.503	0.132	-3.82	0.0004

The model works, and in the expected direction: union participation indexes a decrease, and unemployment an increase, of Alzheimer's
mortality incidence in the young elderly, consistent with a theoretical model for which locus-of-control affects HPA axis activity and the
de-facto rate of aging.

It is possible to parse these results further: The Southern US states
form the core of the so-called 'right-to-work' (RTW) laws that forbid

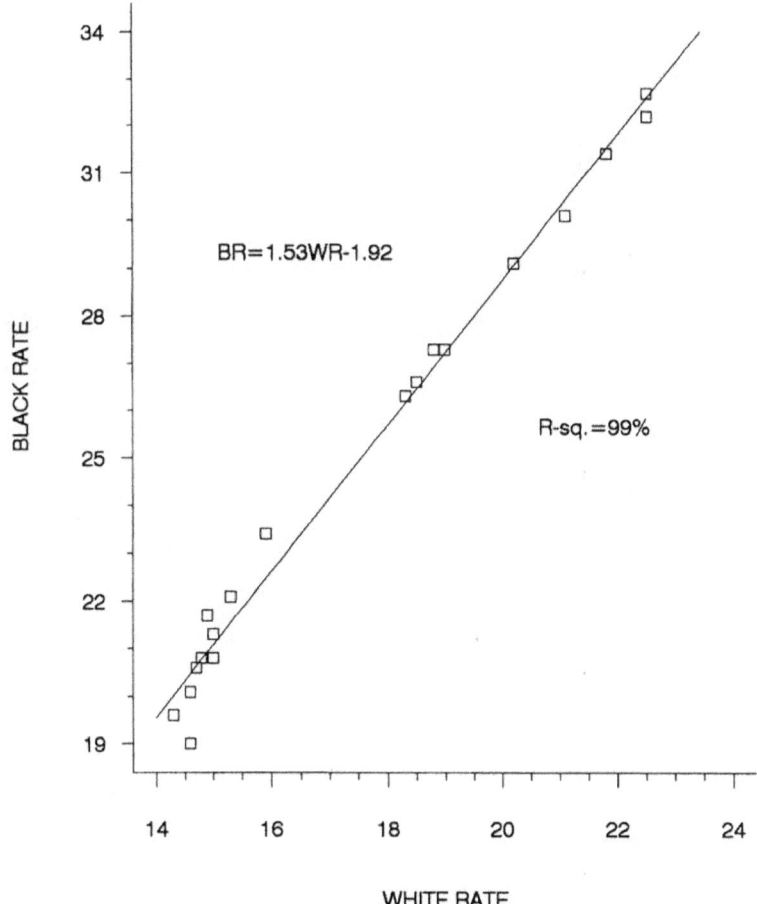

Figure 6.7: US Black vs. White diabetes death rates (per 10^5), 1979-97. While the Black rates are uniformly higher, strong coupling implies both populations are closely linked in a deteriorating social structure.

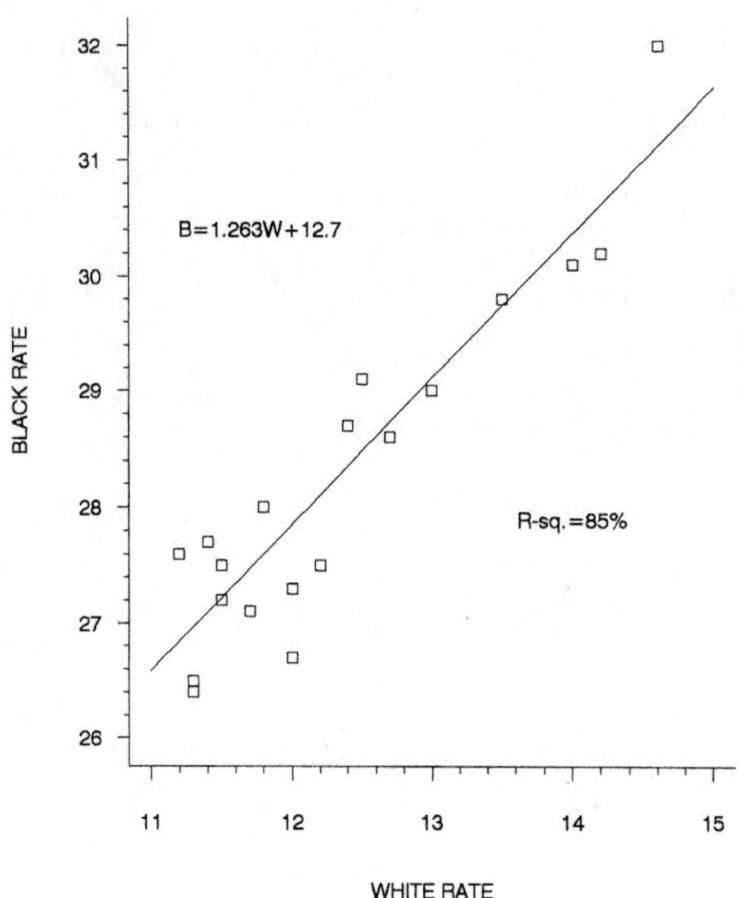

Figure 6.8: Same as figure 6.7 for US hypertension death rates.

requiring a worker to join a union even if he or she is employed in a work force that has union representation. RTW laws indicate and instantiate a culture of individualism and of active anti-collectivism. Other differences between the RTW and non-RTW states include economic history with the former relatively recently industrialized and historically agricultural. Indeed, the plains and Southwestern RTW states showed the fastest rate of increase in manufacturing jobs in the 1990's, but very low numbers of such jobs per unit population. Typically, indicators of social and political engagement such as voting participation and percent employed belonging to a union showed large differences between the two groups of states. Together, they likely indicate the strength of social and political support and control within the two groups of states.

We first examine annual average rates of Alzheimer's deaths per 100,000 over the years 1999-2006 for three age cohorts: 65-74, 75-84, and 85+ across the two sets of states, and compare them using a standard t-test.

Cohort	1	2	3
RTW	23.0	182.2	843.1
non-RTW	19.3	159.3	802.7
P(t-test)	0.02	0.04	NS

The rates, as expected, increase sharply with cohort age, but are markedly lower at all ages in the non-RTW states, and statistically significantly so in the younger.

The regression model above can be applied to all three cohorts and for the US, RTW, and non-RTW states. The percent of adjusted variance accounted for by the models, R^2, and the maximum significance P of the regressions, is as follows:

Cohort	1	2	3	Max. P
US	22.1	33.0	8.7	0.04
non-RTW	16.6	31.8	20.8	0.04
RTW	0	12.4	0	NS

Although the RTW regressions are not significant, the others are highly so, and the raw numbers all show a peak in the middle cohort.

Figure 6.9 displays these results as a signal-to-noise ratio vs. signal amplitude graph. Age is taken as a de-facto toxic exposure, and the regression adjusted R^2 as the SN measure for the model.

Two essential points are the failure of the model for the RTW states and the relative shift of the non-RTW toward the older cohort relative to the full US model. The inference is that collective efficacy appears to lower an HPA axis-induced premature aging that expresses itself in

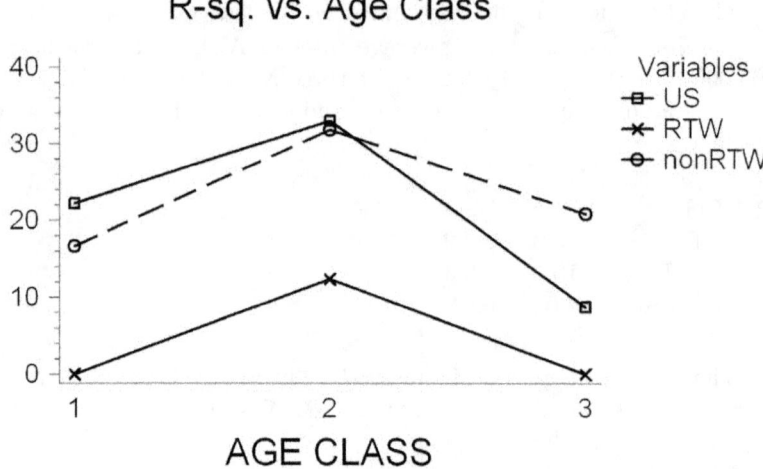

Figure 6.9: Regression adjusted R^2 as signal-to-noise ratio vs. age cohort as toxic exposure for a model based on percent unemployed and percent of workforce unionized. The SNR follows a unimodal 'inverted U' pattern consistent with signal transduction, with the non-RTW states shifted to higher ages. This is consistent with a biosemiotics perspective in which the degree of physiological meaning of the population-level response to the driving variates is changed by the underlying cultural milieu.

Alzheimer's mortality in the young elderly, in accordance with theory. Underlying cultural context appears to profoundly affect both the rate of 'effective aging' and the population response to patterns of affordance and stress.

Further work in this direction should be done using data at the county level of analysis – encompassing some 3,800 distinct geographic entities – to increase the analytic power, illuminate mechanisms at different scales, and explore how policy affects linkages across scales.

A possible inference of these results, partial as they may be, is that Stern's (2009) idea of cognitive reserve may extend to the embedding of the individual in community. That is, in the sense of Wallace and Fullilove (2008), the 'natural' human state includes a powerful measure of distributed cognition that can, in a likely synergistic manner, buffer both cognitive decline and the rate of physiological aging. One implication would be that the RTW states in particular, and the US in general, following the arguments of Heine (2001) and Hecht et al. (2010), embody a pathologically individualistic culture that is contrary to evolved human norms, a disjunction expressing itself in premature aging.

Chapter 7

Anti-public health

Current theory clearly identifies stress as critical to the etiology of visceral obesity, the metabolic syndrome, and their pathological sequelae – including protein folding disorders such as Alzheimer's disease – mediated by the HPA axis and other physiological subsystems.

Both animal and human studies, however, have indicated that not all stressors are equal in their effect: particular forms of domination in animals and lack of control over work activities in humans are well-known to be especially effective in triggering metabolic syndrome and chronic inflammatory coronary lipid deposition. Thus stress can be given meaning from context, as indicated in figures 5.1 and 5.2.

Recent analyses have examined the general association between social status and health in Western subcultures. For example, figure 7.1, from Singe-Manoux et al. (2003), displays a clear dose-response relation between age adjusted prevalence of self-reported ill-health versus self-reported status rank for white collar workers in the UK. 1 is high and 10 low rank. The low status group approaches the 'LD-50' level at which half the population shows a response to dosage.

For the US, figure 7.2 shows the percent of income concentrated in the top five percent of the population as a function of the integral of manufacturing job loss, 1980-1998. The correlation is very high indeed, suggesting that the destruction of the US industrial base consequent on the catastrophic diversion of scientific and engineering resources into the Cold War (e.g., Ullmann, 1988; Wallace, et al., 1999) has had the effect of concentrating wealth and power in the hands of a very small segment of the population. The loss of unionized industrial jobs, and their guarantees of job security, health insurance, retirement benefits, and the exercise of collective power is, in our view, a principal source of a widening population level stress that has been concentrated in African-American populations via the familiar pattern of 'last hired, first fired'.

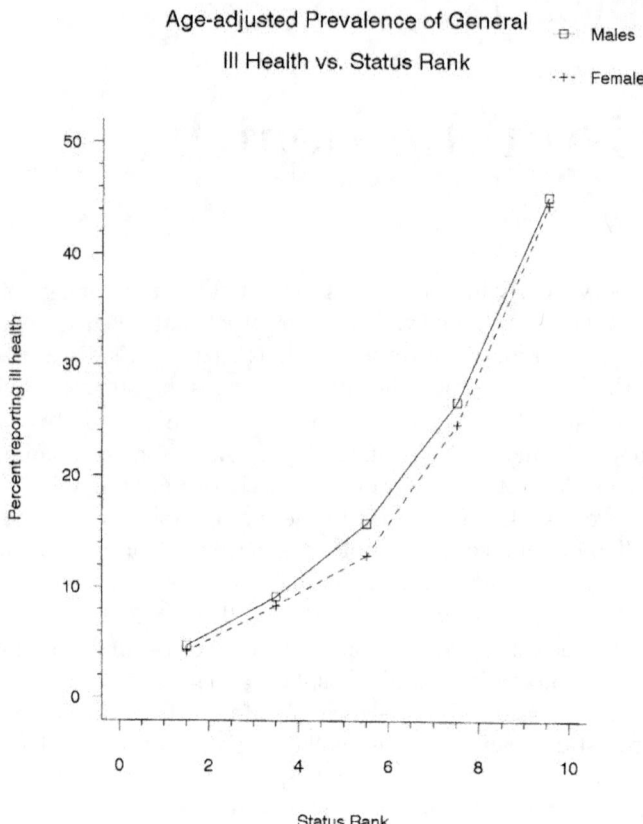

Figure 7.1: Redisplay of data from Singh-Manoux et al. (2003). Sex-specific dose-response curves of age-adjusted self-reported ill-health vs. self-reported status rank, Whitehall II cohort, 1997 and 1999. 1 is high, 10 is low, status. The curve is approaching the LD-50 at which half the dosed population suffers physiological effect of a poison.

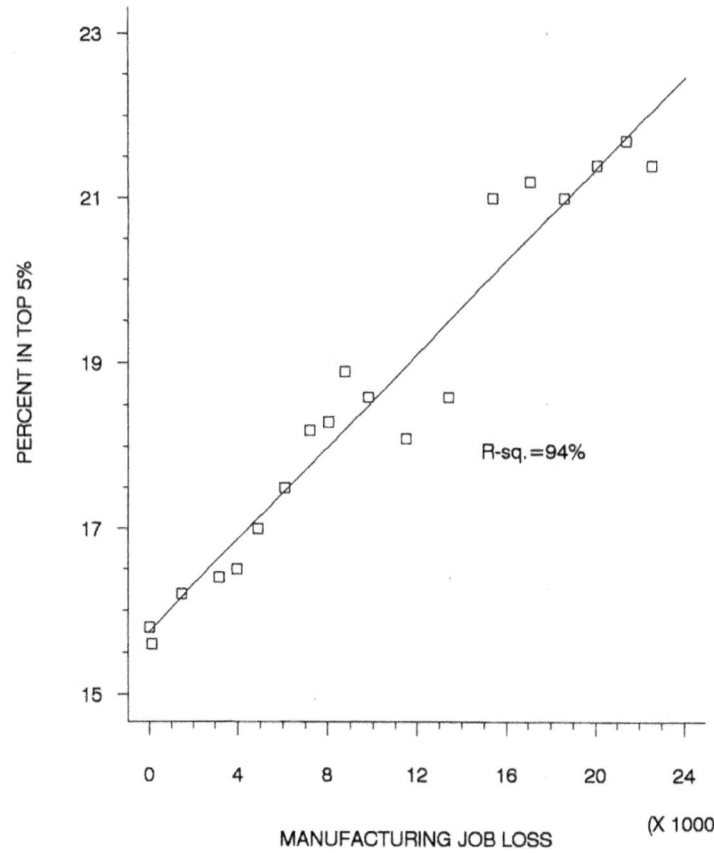

Figure 7.2: Percent of total US income concentrated in the highest 5 percent as a function of the integral of manufacturing job loss, 1980-1998. Devastation of the manufacturing sector consequent on the disruptions caused by the Cold War diversion of engineering and scientific resources from the civilian economy has disempowered vast sections of the US population.

Figure 7.3 shows the simple linear correlation between the annual percent of the US voting age population convicted of a felony between 1980 and 1998 and the integral of manufacturing job loss for the period. The correlation is very good indeed. The percent of felons tripled, serving as yet another index of, and significant contributor to, population level stress.

Our analysis has been in terms of a cognitive HPA axis responding in a culturally-specific manner to a highly structured 'language' of psychosocial stress. We see stress as literally writing a distorted image of itself onto the behavior of the HPA axis in a way analogous to learning plateaus in a neural network or to punctuated equilibrium in a simple evolutionary process. The first form of phase transition/generalized symmetry change might be regarded as representing the progression of a normally 'staged' disease. The other could describe certain pathologies characterized by stasis or only slight change, with staging a rare (and perhaps fatal) event. The works of Barker and his group (e.g., Barker, 2002; Barker et al., 2002) suggests that such HPA axis dysfunction in a mother can become a strong epigenetic catalyst for her children.

To reiterate, psychosocial stress is, for humans, as we have discussed, a culturally specific artifact, one of many such that interact intimately with human physiology. Indeed, much current theory in evolutionary anthropology focuses on the essential (but not unique) role culture plays in human biology (Avital and Jablonka, 2000; Durham, 1991).

If, as the evolutionary anthropologist Robert Boyd has suggested, 'Culture is as much a part of human biology as the enamel on our teeth', what does the rising tide of obesity in the US suggest about American culture and the American system? About 22% of both African-American and Hispanic children are overweight, as compared to about 12% of non-Hispanic whites, and that prevalence is rising across the board (Strauss and Pollack, 2001). This suggests that, while the effects of an accelerating social pathology related to deindustrialization, deurbanization, and loss of democracy may be most severe for ethnic minorities in the US, the larger, embedding, cultural dysfunction has already spread upward along the social hierarchy, and is quickly entraining the majority population as well. This will, according to our analysis, become expressed in rising rates of protein folding and related disorders for majority and minority populations, although not at the same rates.

This is an explanation whose policy implications stand in stark contrast to current individual-oriented exhortations about 'taking responsibility for one's behavior' or 'eating less and getting more exercise' (Hill et al., 2003). The US liberal approach is to mirror the explanations of the failed drug war: People overeat because there's a McDonald's on

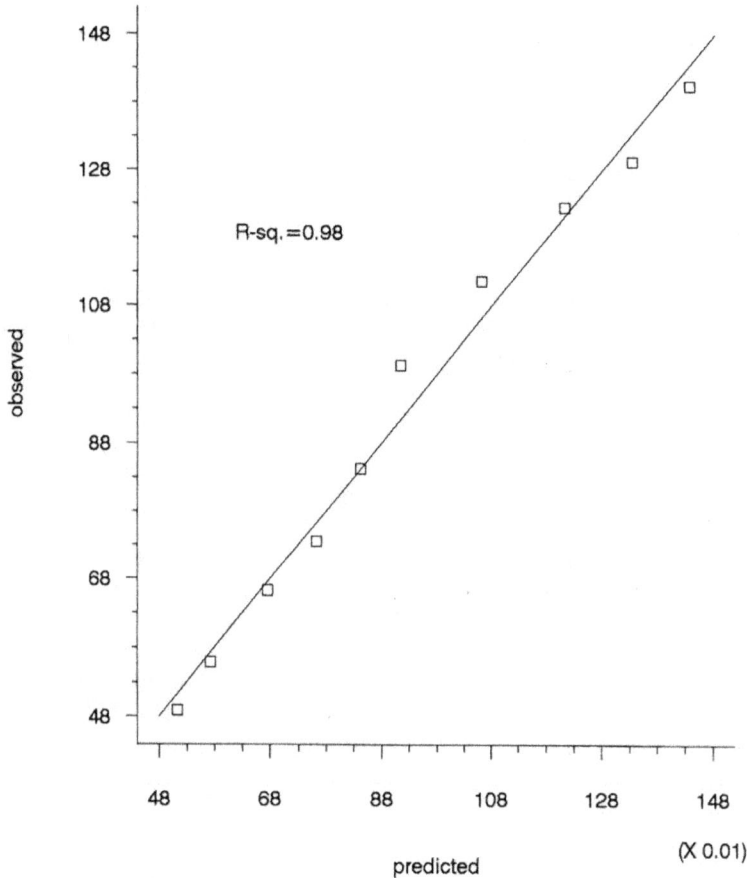

FELONS AS % OF VOTING-AGE POPULATION VS.
INTEGRAL OF MANUFACTURING JOB LOSS 80-98
(X 0.01)

Figure 7.3: Felons as a percent of the US voting-age population, 1980-1998, expressed as a linear function of the integral of the manufacturing job loss over the period. The proportion of felons tripled. The high correlation suggests that the loss of stable, unionized, manufacturing jobs in the US affects public order as well as public health, and these deteriorations, of course, are likely to be powerfully synergistic.

every street corner, companies market bigger portions, the food they sell is fatty, and so on. In contrast, we find that the fundamental cause of the obesity epidemic over the last twenty years is not television, the automobile, or junk food. All were significant features of American life from the late 1950's into the 1980's without an obesity epidemic. The fundamental cause of the US obesity epidemic is a massive threat to the population caused by continuing deterioration of basic US social, economic, and related structures, in the particular context of a ratcheting of dominance relations resulting from the concentration of effective power within a shrinking elite. This phenomenon is literally writing a life-threatening image of itself onto the bodies of American adults and children through an array of illnesses, including protein folding disorders such as Alzheimer's and Parkinson's diseases. There is a large and growing literature on other aspects of the sharpening inequalities within the US system (particularly Wilkinson, 1996, and related material), and our conclusions fit within that body of work.

The basic and highly plieotropic nature of the biological relation between structured psychosocial stress and cognitive physiological systems ensures that 'magic bullet' interventions will be largely circumvented: in the presence of a continuing socioeconomic and political ratchet, 'medical' modalities are likely to provide little more than the equivalent of a choice of death by hanging or firing squad.

Effective intervention against obesity and its pathological sequelae in the US – including a wide spectrum of protein folding disorders – is predicated on creation of a broad, multi-level, ecological control program. It is evident that such a program must include redress of the power relations between groups, rebuilding of urban (and, increasingly, suburban) minority communities, and effective reindustrialization. This implies the necessity of a resurgence of the labor union, religious, civil rights, and community-based political activities which have been traditionally directed against cultural patterns of injustice in the past, activities which, ultimately, liberate all.

Some final thoughts:

The fidelity of the translation between genome and final protein conformation can be characterized by a rate distortion function, evolutionarily sculpted in the sense of Onuchic and Wolynes (2004), that determines possible phase transitions defining different degrees of protein symmetry.

The protein folding funnel follows a spontaneous symmetry breaking mechanism with average distortion as the temperature analog. In the developmental picture, a dual perspective, the rate distortion function serves as the 'temperature' so that increased channel capacity leads more directly to the final state S_f. The resulting equivalence class groupoid should be similar to the classifications of figure 1.1, extended by the considerations of Section 4.1 regarding intrinsically disordered proteins.

The various outcomes to *in vivo* protein folding – normal, corrected, eliminated, pathological – emerge, in the expanded 'Onsager equation' statistical model of Chapter 3, based on a cognitive paradigm for the regulation of the process, as distinct 'resilience' modes of a complicated internal cellular ecosystem. These are subject to punctuated transitions driven, in some cases, by structured signals from embedding epigenetic and ecological information sources. Increase in the rate of folding disorders with age emerges through a long-time generalization of the Onsager model in which, for humans, culturally-specific – and often policy-driven – excitation of the HPA axis can serve as a cause of both premature aging and, via a generalized retina argument, of therapeutic failure.

In essence, this work extends Tlusty's (2010a) elegant topological exploration of the evolution of the genetic code, suggesting that rate distortion considerations are central to a broad spectrum of molecular biological phenomena, although different measures may come to the fore under different perspectives.

The *in vivo* cognitive paradigm introduced here opens a unified biological vision of protein folding and its disorders that may relate the etiology of a large set of common misfolding and aggregation diseases more clearly to both cellular and epigenetic processes and culturally-mediated environmental stressors (e.g., Schnabel, 2010). This would be, in the current reductionist sandstorm (e.g., Kolata, 2010), no small thing. A cognitive paradigm subsumes epigenetic and environmental catalysis of protein conformation 'development' within a single grammar and syntax, and allows both normal folding and its pathologies to both be viewed as 'natural' outcomes, a perspective more consistent with rates of folding and aggregation disorders observed within an aging population.

Such a cognitive paradigm, as we have constructed it, will likely serve as the foundation for a new class of statistical tools – based on the asymptotic limit theorems of information theory rather than on the Central Limit Theorem alone – that should be useful in the analysis of data related to protein misfolding and aggregation disorders.

We have, in the sense of Heine (2001) and Wallace (2007), focused on the larger physiological and, for humans, cultural, contexts of protein folding and its disorders, as discomforting as these may be for Americans. These contexts include epigenetic and life history stress factors related to large-scale public policies that can act as catalysts to induce highly structured 'large deviations' that accelerate the deterioration of protein folding regulation from the individual to the population level. And we have done this from the ground up, as it were, providing a 'basic biological' model of what Qui et al. (2009), Wilson et al. (2005), and others have observed. A narrow focus on medical magic bullets

(Kolata, 2010; O'Connor et al., 2010) is not consonant with the wide scale of protein folding regulation and its dysfunctions, and a successful search for effective interventions will necessarily involve far broader perspectives than seem comfortable to the strongly culture-bound majority of senior American researchers and the policymaking elites that employ them.

Chapter 8

Mathematical Appendix

8.1 Groupoids

Following Weinstein (1996) closely, a groupoid, G, is defined by a base set A upon which some mapping – a morphism – can be defined. Note that not all possible pairs of states (a_j, a_k) in the base set A can be connected by such a morphism. Those that can define the groupoid element, a morphism $g = (a_j, a_k)$ having the natural inverse $g^{-1} = (a_k, a_j)$. Given such a pairing, it is possible to define 'natural' end-point maps $\alpha(g) = a_j, \beta(g) = a_k$ from the set of morphisms G into A, and a formally associative product in the groupoid $g_1 g_2$ provided $\alpha(g_1 g_2) = \alpha(g_1), \beta(g_1 g_2) = \beta(g_2)$, and $\beta(g_1) = \alpha(g_2)$. Then the product is defined, and associative, $(g_1 g_2) g_3 = g_1(g_2 g_3)$.

In addition, there are natural left and right identity elements λ_g, ρ_g such that $\lambda_g g = g = g \rho_g$ (Weinstein, 1996).

An orbit of the groupoid G over A is an equivalence class for the relation $a_j \sim G a_k$ if and only if there is a groupoid element g with $\alpha(g) = a_j$ and $\beta(g) = a_k$. Following Cannas da Silva and Weinstein (1999), we note that a groupoid is called transitive if it has just one orbit. The transitive groupoids are the building blocks of groupoids in that there is a natural decomposition of the base space of a general groupoid into orbits. Over each orbit there is a transitive groupoid, and the disjoint union of these transitive groupoids is the original groupoid. Conversely, the disjoint union of groupoids is itself a groupoid.

The isotropy group of $a \in X$ consists of those g in G with $\alpha(g) = a = \beta(g)$. These groups prove fundamental to classifying groupoids.

If G is any groupoid over A, the map $(\alpha, \beta) : G \to A \times A$ is a morphism from G to the pair groupoid of A. The image of (α, β) is the orbit equivalence relation $\sim G$, and the functional kernel is the union of the isotropy groups. If $f : X \to Y$ is a function, then the kernel of f, $ker(f) = [(x_1, x_2) \in X \times X : f(x_1) = f(x_2)]$ defines an equivalence relation.

Groupoids may have additional structure. As Weinstein (1996) explains, a groupoid G is a topological groupoid over a base space X if G and X are topological spaces and α, β and multiplication are continuous maps. A criticism sometimes applied to groupoid theory is that their classification up to isomorphism is nothing other than the classification of equivalence relations via the orbit equivalence relation and groups via the isotropy groups. The imposition of a compatible topological structure produces a nontrivial interaction between the two structures. Below we will introduce a metric structure on manifolds of related information sources, producing such interaction.

In essence, a groupoid is a category in which all morphisms have an inverse, here defined in terms of connection to a base point by a meaningful path of an information source dual to a cognitive process.

As Weinstein (1996) points out, the morphism (α, β) suggests another way of looking at groupoids. A groupoid over A identifies not only which elements of A are equivalent to one another (isomorphic), but *it also parametrizes the different ways (isomorphisms) in which two elements can be equivalent*, i.e., all possible information sources dual to some cognitive process. Given the information theoretic characterization of cognition presented above, this produces a full modular cognitive network in a highly natural manner.

Brown (1987) describes the fundamental structure as follows:

> A groupoid should be thought of as a group with many objects, or with many identities... A groupoid with one object is essentially just a group. So the notion of groupoid is an extension of that of groups. It gives an additional convenience, flexibility and range of applications...
>
> EXAMPLE 1. A disjoint union [of groups] $G = \cup_\lambda G_\lambda, \lambda \in \Lambda$, is a groupoid: the product ab is defined if and only if a, b belong to the same G_λ, and ab is then just the product in the group G_λ. There is an identity 1_λ for each $\lambda \in \Lambda$. The maps α, β coincide and map G_λ to $\lambda, \lambda \in \Lambda$.
>
> EXAMPLE 2. An equivalence relation R on [a set] X becomes a groupoid with $\alpha, \beta : R \to X$ the two projections, and product $(x, y)(y, z) = (x, z)$ whenever $(x, y), (y, z) \in R$. There is an identity, namely (x, x), for each $x \in X$...

Weinstein (1996) makes the following fundamental point:

> Almost every interesting equivalence relation on a space B arises in a natural way as the orbit equivalence relation of some groupoid G over B. Instead of dealing directly with the orbit space B/G as an object in the category S_{map} of sets and mappings, one should consider instead

the groupoid G itself as an object in the category G_{htp} of groupoids and homotopy classes of morphisms.

The groupoid approach has become quite popular in the study of networks of coupled dynamical systems which can be defined by differential equation models, (e.g., Golubitsky and Stewart 2006).

8.2 Stochastic differential equations

8.2.1 Martingales

Suppose we have entered one of the great gambling casinos of the world, host to an almost infinite variety of games of chance: card games ranging from baccarat and blackjack to keno and poker, roulette wheels, one armed-bandits, dice games, and so on. Each game has different rules of play, even if, as for card games, the instruments of play are all the same. Complicated outcomes for those instruments produce equally complex patterns of loss or gain for the player.

Suppose a player begins with an initial fortune of some given amount, and bets $n = 1, 2, ...$ times according to a stochastic process in which a stochastic variable $\mathbf{X}_n$, which represents the size of the player's fortune at play n, takes values $\mathbf{X}_n = x_{n,i}$ with probabilities $P_{n,i}$ such that $\sum_i P_{n,i} = 1$, where i represents a particular outcome at step n.

Assume for all n there exists a value $0 < C < \infty$ such that the expectation of $\mathbf{X}_n$,

$$E(\mathbf{X}_n) \equiv \sum_i x_{n,i} P_{n,i} < C$$

(8.1)

for all n. That is, no infinite or endlessly increasing fortunes are permitted.

We note that the state $\mathbf{X}_n = 0$, having probability P_n^0, i.e., the loss of all a player's funds, terminates the game.

We suppose it possible to define conditional probabilities at step $n + 1$ which depend on the way in which the value of $\mathbf{X}_n$ was reached, so that we can define the conditional expectation of $\mathbf{X}_{n+1}$:

$$E(\mathbf{X}_{n+1}|\mathbf{X}_1, \mathbf{X}_2, ...\mathbf{X}_n) \equiv E(\mathbf{X}_{n+1}|n).$$

The 'sample space' for the probabilities defining this conditional expectation is the set of different possible sequences of the $x_{m,i} > 0$ such that $x_{1,i}, x_{2,j}, x_{3,k}...x_{n,q}$.

We call the sequence of stochastic variables $\mathbf{X}_n$ defining the game a *Submartingale* if, at each step n, $E(\mathbf{X}_{n+1}|n) \geq \mathbf{X}_n$, a *Martingale* if

$$E(\mathbf{X}_{n+1}|n) = \mathbf{X}_n$$

and a *Supermartingale* if

$$E(\mathbf{X}_{n+1}|n) \leq \mathbf{X}_n$$

$\mathbf{X}_n$ is, remember, the player's fortune at step n.

Clearly a submartingale is favorable to the player, a martingale is an absolutely fair game, and a supermartingale is favorable to the house.

Regardless of the complexity of the game, the details of the playing instruments, the ways of determining gains or loss or their amounts, or any other structural factors of the underlying stochastic process, the essential content of the Martingale Limit Theorem is that in all three cases the sequence of stochastic variables $\mathbf{X}_n$ converges in probability 'almost everywhere' to a well-defined stochastic variable $\mathbf{X}$ as $n \to \infty$. That is, for each kind of martingale, no matter the actual sequence of winnings $x_{1,i}, x_{2,j}, ...x_{n,k}, x_{n+1,m}, ...$, you get to the same limiting stochastic variable $\mathbf{X}$. Sequences for which this does not happen have zero probability.

A simple proof of this result (Petersen, 1995) runs to several pages of dense mathematics using modern theories of abstract integration on sets. Indeed, all the asymptotic theorems we have cited require more or less arduous application of measure theory and Lebesgue integration, topics which are themselves relatively straightforward, elegant and worth study (Rudin, 1976, Royden, 1968). Proofs using more elementary approaches (Karlin and Taylor, 1975, Ch. 6) run to full chapters.

8.2.2   Nested Martingales

We are interested in a compound stochastic process in which the 'winnings' at the 'smaller' scale, played by one set of rules, contribute, in some sense, to a quite different game having completely different rules on a 'larger' scale. These games are bounded by the condition $E(\mathbf{X}_n) < C$, for some finite positive C.

The essential point is that a proportion of the winnings from the smaller game are duplicated by a 'benefactor' and directly raise the magnitude of the player's fortune for the larger, embedding game.

If the inner game is characterized at step n by the random variable $\mathbf{Y}_n$, then the 'real' winnings at step $n + 1$ for the embedding game, associated with the random variable $\mathbf{X}_{n+1}$, become, for some function f_n, which may involve additional stochastic variables,

$$\mathbf{X}_{n+1} = f_n(\mathbf{X}_n, \mathbf{Y}_n, \mathbf{Y}_{n+1}).$$

(8.2)

A slightly different approach would involve conditional expectations in the convolution of scales:

$$E(\mathbf{X}_{n+1}|n) = F_n(\mathbf{X}_n, \mathbf{Y}_n, E(\mathbf{Y}_{n+1}|n))$$

(8.3)

for some function F_n.

Traditionally, the simplest version of this extension assumes that the compound game is, in some sense, a subset of the original:

$$\mathbf{X}_{n+1} = \mathbf{X}_n + \mathbf{A}_n(\mathbf{Y}_{n+1} - \mathbf{Y}_n).$$

(8.4)

We assume the filter $\mathbf{A}_n \geq 0$ is a non-negative stochastic variable, which can indeed take the value 0. This may, for example, be greater than zero only one time in ten or a hundred, on average. Taking the conditional expectation gives

$$E(\mathbf{X}_{n+1}|n) = \mathbf{X}_n + \mathbf{A}_n(E(\mathbf{Y}_{n+1}|n) - \mathbf{Y}_n)$$

(8.5)

where we recognize the conditional expectation of any variate $\mathbf{Z}_n$ at step n is just its value.

Since $\mathbf{A}_n \geq 0$, *the game described by the attenuated sequence* $\mathbf{X}_n$ *has the same martingale classification as does the nested central city game described by* $\mathbf{Y}_n$.

8.2.3   The Martingale Transform

The X-processes in equation (8.4) is the *Martingale transform* of $\mathbf{Y}_n$ (Taylor, 1996, p.232; Billingsley, 1968, p. 412), and the result is classic, representing the *impossibility of a successful betting system*.

Note that the basic Martingale transform can be rewritten as

$$\frac{\mathbf{X}_{n+1} - \mathbf{X}_n}{\mathbf{Y}_{n+1} - \mathbf{Y}_n} \equiv \frac{\Delta \mathbf{X}_n}{\Delta \mathbf{Y}_n} = \mathbf{A}_n,$$

or

$$\Delta \mathbf{X}_n = \mathbf{A}_n \Delta \mathbf{Y}_n.$$

(8.6)

Induction gives

$$\mathbf{X}_{n+1} = \mathbf{X}_0 + \sum_{j=1}^{n} \mathbf{A}_j \Delta \mathbf{Y}_j.$$

(8.7)

This notation is suggestive: in fact the Martingale transform is the discrete analog of Ito's stochastic integral relative to a sequence of stopping times, (Taylor, 1996, p. 232; Protter, 1990, p. 44; Ikeda and Watanabe, 1989, p. 48). In the stochastic integral context the Y-process is called the 'integrator' and the A-process the 'integrand.' Further development leads toward generalizations of Brownian motion, the Poisson process, and so on (Meyer, 1989; Protter, 1990).

The basic picture is of the transmission of a signal, $\mathbf{Y}_n$, in the presence of noise, $\mathbf{A}_n$.

8.2.4 Stochastic Differential Equations

A more realistic extension of the elementary denumerable Martingale transform for our purposes is

$$\mathbf{X}_{n+1} = \mathbf{X}_n + (\mathbf{B}_{n+1} - \mathbf{B}_n)\mathbf{X}_n + \mathbf{A}_n(\mathbf{Y}_{n+1} - \mathbf{Y}_n),$$

(8.8)

where $\mathbf{B}_n$ is another stochastic variable.

Using the more suggestive notation of equations (8.6) and (8.7) this becomes the fundamental stochastic differential equation

$$\Delta\mathbf{X}_n = \mathbf{X}_n\Delta\mathbf{B}_n + \mathbf{A}_n\Delta\mathbf{Y}_n.$$

(8.9)

Taking conditional expectations gives

$$E(\mathbf{X}_{n+1}|n) - \mathbf{X}_n =$$

$$\mathbf{X}_n(E(\mathbf{B}_{n+1}|n) - \mathbf{B}_n) + \mathbf{A}_n(E(\mathbf{Y}_{n+1}|n) - \mathbf{Y}_n).$$

(8.10)

If $\mathbf{X}_n, \mathbf{A}_n \geq 0$, the martingale classification of $\mathbf{X}$ depends on those of $\mathbf{B}$ and $\mathbf{Y}$.

Extending the argument to a hierarchically-linked network is straightforward, leading to the Ito stochastic integral

$$\mathbf{X}_{n+1} \approx \mathbf{X}_0 + \sum_{k=1}^{n} \mathbf{A}_k\Delta\mathbf{Y}_k.$$

(8.11)

The complete hierarchical system, then undergoes an iterative Z-process defined by the integrator $\mathbf{X}_j$:

$$\mathbf{Z}_{m+1} \approx \mathbf{Z}_0 + \sum_{j=1}^{m} \mathbf{C}_j \Delta \mathbf{X}_j.$$

(8.12)

Extension of this development to intermediate times is complicated and involves taking the continuous limit of the Riemann-type sums of equations (8.7), (8.11) and (8.12). This produces the stochastic differential equation

$$d\mathbf{X}_t = \mathbf{X}_t d\mathbf{B}_t + \mathbf{A}_t d\mathbf{Y}_t,$$

(8.13)

whose solution depends critically on the behavior of the second-order step-by-step 'quadratic variation,' a variance-like limit of the stochastic processes. Letting $\mathbf{U}_n, \mathbf{V}_n$ be two arbitrary processes with $\mathbf{U}_0 = \mathbf{V}_0 = 0$, their quadratic variation is

$$[\mathbf{U}_n, \mathbf{V}_n] \equiv \sum_{j=1}^{n-1} (\mathbf{U}_{j+1} - \mathbf{U}_j)(\mathbf{V}_{j+1} - \mathbf{V}_j).$$

(8.14)

Taking the 'infinitesimal limit' of continuous time, a term-by-term expansion of this sum can be shown to give (e.g., Meyer, 1989; Protter, 1990)

$$[\mathbf{U}_t, \mathbf{V}_t] = \mathbf{U}_t\mathbf{V}_t - \int_0^t \mathbf{U}_s d\mathbf{V}_s - \int_0^t \mathbf{V}_r d\mathbf{U}_r.$$

(8.15)

To put this in some perspective, classical Brownian motion has the 'structure equation' $[\mathbf{X}_t, \mathbf{X}_t] = t$.

That is, for Brownian motion the jump-by-jump quadratic variation increases linearly with time. While much of the contemporary theory of financial markets is based on Brownian analogs, real processes are likely to be more complex, subject to sudden, massive, discontinuous 'phase changes' which cannot be simply characterized as diffusional.

The solution of equation (8.13) is a classic result in the theory of stochastic differential equations (Protter, 1990). We assume for simplicity no discontinuous jumps, and first study the 'exponential' equation

$$d\mathbf{X}_t = \mathbf{X}_t d\mathbf{B}_t$$

or equivalently

$$\mathbf{X}_t = \mathbf{X}_0 + \int_0^t \mathbf{X}_s d\mathbf{B}_s$$

(8.16)

Following Protter (1990, p. 78) this has the solution

$$\mathbf{X}_t = \epsilon(\mathbf{B})_t = \mathbf{X}_0 \exp(\mathbf{B}_t - 1/2[\mathbf{B}_t, \mathbf{B}_t]).$$

(8.17)

Next we define

$$\mathbf{H}_t \equiv \int_0^t \mathbf{A}_s d\mathbf{Y}_s.$$

(8.18)

Equation (8.13) can be restated as

$$\mathbf{X}_t = \mathbf{H}_t + \mathbf{X}_0 + \int_0^t \mathbf{X}_s d\mathbf{B}_s.$$

(8.19)

For the continuous case, this has the formal solution (Protter, 1990, p.266)

$$\epsilon_{\mathbf{H}}(\mathbf{B})_t =$$

$$\epsilon(\mathbf{B})_t [\mathbf{H}_0 + \int_0^t 1/\epsilon(\mathbf{B})_s d(\mathbf{H}_s - [\mathbf{H}, \mathbf{B}]_s)],$$

(8.20)

with $1/\epsilon(\mathbf{B}) = \epsilon(-\mathbf{B} + [\mathbf{B}, \mathbf{B}])$.

The structure equations defining $[\mathbf{B}, \mathbf{B}]$ and $[\mathbf{H}, \mathbf{B}]$ are critical in determining transient behavior, but not likely to have simple Brownian form.

Chapter 9

References

Alzheimer's Association, 2006, African-Americans and Alzheimer's disease: the silent epidemic. Available for download from www. alz.org.

Andre, I., C. Strauss, D. Kaplan, P. Bradley, and D. Baker, 2008, Emergence of symmetry in homooligomeric biological assemblies, *Proceedings of the National Academy of Sciences*, 105:16148-16152.

Anfinsen, C., 1973, Principles that govern the folding of protein chains, *Science*, 181:223-230.

Ash, R., 1990, *Information Theory*, Dover, New York.

Astbury, W., 1935, The x-ray interpretation of the denaturation and the structure of the seed gobulins, *Biochemistry*, 29:2351-2360.

Atlan, H., and I. Cohen, 1998, Immune information, self-organization, and meaning, *International Immunology*, 10:711-717.

Avital E. and E. Jabolnka, 2000, *Animal Traditions: Behaviroal Inheritance in Evolution*, Cambridge University Press, UK.

Beck, C., and F. Schlogl, 1995, *Thermodynamics of Chaotic Systems*, Cambridge University Press, New York.

Barker D., 2002, Fetal programming of coronary heart disease, *Trends in Endocrinology and Metabolism*, 13:364-372.

Barker D., T. Forsen, A. Uutela, C. Osmond, and J. Erikson, 2002, Size at birth and resistance to effects of poor living conditions in adult life: longitudinal study, *British Medical Journal*, 323:1261-1262.

Barkow, J., L. Cosmides, J. Tooby (eds.), 1992, *The Adapted Mind: Biological Approaches to Mind and Culture*, University of Toronto Press.

Bedekar, A., 2001, On the information about message arrival times required for in-order decoding, in *Proceedings of the International Symposium on Information Theory (ISIT)*, Washington, D.C. 2201:227.

Bennett, C., 1988, Logical depth and physical complexity. In Herkin, R. (ed.), *The Universal Turing Machine: A Half-Century Survey*, Oxford University Press, pp. 227-257.

Billingsley, P., 1968, *Convergence of Probability Measures*, John Wiley and Sons, New York.

Bos, R., 2007, Continuous representations of groupoids. arXiv:math/0612639.

Brown, R., 1987, From groups to groupoids: a brief survey, *Bulletin of the London Mathematical Society*, 19:113-134.

Bruce, M., and A. Dickinson, 1987, Biological evidence that scrapie agent has an independent genome, *Journal of General Virology*, 68:79-89.

Buneci, M., 2003, *Representare de Groupoizi*, Editura Mirton, Timosoara, Romania.

Cannas Da Silva, A., and Weinstein, A., 1999, *Geometric Models for Noncommutative Algebras*, American Mathematical Society, Providence, RI.

CDC, 2003, 1991-2001 prevalence of obesity among US adults http://www.cdc.gov/nccdchp/dnpa/obesity/trend/prev$_c$har.htm

Champagnat, N., R. Ferriere, and S. Meleard, 2006, Unifying evolutionary dynamics: From individual stochastic processes to macroscopic models, *Theoretical Population Biology*, 69:297-321.

Chiti, F., P. Webster, N. Taddei, A. Clark, M. Stefani, G. Ramponi, and C. Dobson, 1999, Designing conditions for *in vitro* formation of amyloid protofilaments and fibrils, *Proceedings of the National Academy of Sciences of America*, 96:3590-3594.

Chou, K.C., and L. Carlacci, 1991, Energetic approach to the folding of α/β barrels, *Proteins: Structure, Function and Genetics*, 9:280-295.

Chou, K.C., and C.T. Zhang, 1995, Prediction of protein structural classes, *Reviews in Biochemistry and Molecular Biology*, 30:275-349.

Chou, K.C., and G. Maggiora, 1998, Domain structural class prediction, *Protein engineering*, 11:523-528.

Chou, K., and Y.D. Cai, 2004, Predicting protein structural class by functional domain composition, *Biochemical an Biophysical Research Communications*, 321:1007-1009.

Cohen, I., 2000, *Tending Adam's Garden: Evolving the Cognitive Immune Self*, Academic Press, New York.

Collinge, J., and A. Clarke, 2007, A general model of prion strains and their pathogencity, *Science*, 318:930-936.

Cover, T., and H. Thomas, 1991, *Elements of Information Theory*, Wiley, New York.

de Groot, S., and P. Mazur, 1984, *Nonequilibrium Thermodynamics*, Dover, New York.

Dembo, A., and O. Zeitouni, 1998, *Large Deviations and Applications*, 2nd edition, Springer, New York.

Diekmann U., and R. Law, 1996, The dynamical theory of coevolution: a derivation from stochastic ecological processes, *Journal of*

Mathemaical Biology, 34:579-612.

Dill, K., S. Banu Ozkan, T. Weikl, J. Chodera, and V. Voelz, 2007, The protein folding problem: when will it be solved? *Current Opinion in Structural Biology*, 17:342-346.

Dobson, C., 2003, Protein folding and misfolding, *Nature*, 426:884-890.

Dong, H., Csernansky, J., 2009, Effects of stress and stress hormones on Amyloid-β Protein and plaque deposition, *Journal of Alzheimer's Disease*, 18:459-469.

Durham, W., 1991 *Coevolution: Genes, Culture and Human Diversity*, Stanford University Press, Palo Alto CA.

Duryea, P., 1978, Press release dated Friday, Jan. 27, 1978, Office of the Republican Assembly Leader, Albany, NY.

El Gamal, A., and Y. Kim, 2010, Lecture notes on network information theory, arXiv:1001.3404v4.

Ellis, R., 1985, *Entropy, Large Deviations, and Statistical Mechanics*, Springer, New York.

English, T., 1996, Evaluation of evolutionary and genetic optimizers: no free lunch. In Fogel, L., P. Angeline, and T. Back (eds.), *Evolutionary Programming V: Proceedings of the Fifth Annual Conference on Evolutionary Programming*, 163-169, MIT Press, Cambridge, MA.

Epel, E., E. Blackburn, J. Lin, F. Dhabhar, N. Adler, J. Morrow, R. Cawthon, 2004, Accelerated telomere shortening in response to life stress, *Proceedings of the National Academy of Sciences*, 101:17312-17315.

Falsig, J., K. Nilsson, T. Knowles, and A. Aguzzi, 2008, Chemical and biophysical insights into the propagation of prion strains. *HFSP Journal*, 2:332-341.

Feynman, R., 2000, *Lectures on Computation*, Westview, New York.

Fillit, H., D. Nash, T. Rundek, and A. Zukerman, 2008, *American Journal of Geriatric Pharmacotherapy*, 6:100-118.

Glazebrook, J.F., and R. Wallace, 2009a, Small worlds and red queens in the global workspace: an information-theoretic approach, *Cognitive Systems Research*, 10:333-365,

Glazebrook, J.F., and R. Wallace, 2009b, Rate distortion manifolds as models for cognitive information, *Informatica*, 33:309-345.

Goldschmidt, L., P. Teng, R. Riek, and D. Eisenberg, 2010, Identifying the amylome, proteins capable of forming amyloid-like fibrils, *Proceedings of the National Academy of Sciences*, 107:3487-3492.

Golubitsky M., and I. Stewart, 2006, Nonlinear dynamics and networks: the groupoid formalism, *Bulletin of the American Mathematical Sociey*, 43:305-364.

Goodsell, D., and A. Olson, 2000, Structural symmetry and protein function, *Annual Reviews of Biophysics and Biomolecular Structure*, 29:105-153.

Grubele, M., 2005, Downhill protein folding: evolution meets physics, *Comptes Rendus Biologies*, 328:701-712.

Gunderson, L., 2000, Ecological resilience in theory and application, *Annual Reviews of Ecological Systematics*, 31:425-439.

Gurland, B., D. Wilder, R. Lantigua, Y. Stern, J. Chen, E. Killeffer, and R. Mayeux, 1999, Rates of dementia in three ethnoracial groups, *International Journal of Geriatric Psychiatry*, 14:481-493.

Haataja, L., T. Gurlo, C. Huang, and P. Butler, 2008, Islet amylod in type 2 diabetes, and the toxic oligomer hypothesis, *Endocrine Reviews*, 29:303-316.

Hartl, F., and M. Hayer-Hartl, 2009, Converging concepts of protein folding *in vitro* and *in vivo*, *Nature Structural and Molecular Biology*, 16:574-581.

Hebert, D., and M. Molinari, 2007, Protein folding, quality control, degradation, and related human diseases, *Physiologal Reviews*, 87: 1377-1408.

Hecht, M., A. Das, A. Go, L. B. Aradley, and Y. Wei, 2004, *Protein Science*, 13:1711-1723.

Heine, S., 2001, Self as cultural product: an examination of East Asian and North American selves, *Journal of Personality*, 69:881-906.

Henrich, J., S. Heine, and A. Norenzayan, 2010, The Weirdest people in the world, *Behavioral and Brain Sciences*, 33:61-135.

Hill, J., H. Wyatt, G. Reed, J. Peters, 2003, Obesity and the environment: where do we go from here? *Science*, 266:853-858.

Holling, C., 1973, Resilience and stability of ecological systems, *Annual Reviews of Ecological Systematics*, 4:1-23.

Huang, Y., Z. Liu, 2009, Kinetic advantage of intrinsically disordered proteins in coupled folding-binding process: a critical assessment of the 'fly-casting' mechanism, *Journal of Molecular Biology*, 393:1143-1159.

Ikeda, N., and S. Watanabe, 1989, *Stochastic Differential Equations and Diffusion Processes*, second edition, North Holland Publishing Co., Amsterdam.

Ives, A., 1995, Measuring resilience in stochastic systems, *Ecological Monographs*, 65:217-233.

Ivankov, D., S. Garbuzynsky, E. Alm, K. Plaxco, D. Baker, and A. Finkelstein, 2003, Contact order revisited: influence of protein size on the folding rate, *Protein Science*, 12:2057-2062.

Kamtekar, S., J. Schiffer, H. Xiong, J. Babik, and M. Hechtg, 1993, Protein deisgn by patterning of polar and nonpolar amino acids. *Science*, 262:1680-1685.

Karlin, S., and H. Taylor, 1975, *A First Course in Stochastic Processes*, second edition, Academic Press, New York.

Khinchin, A., 1957, *Mathematical Foundations of Information Theory*, Dover, New York.

Kim, W., and M. Hecht, 2006, Generic hydrophobic residues are sufficient to promote aggregation of the Alzheimer's Aβ42 peptide, *Proceedings of the National Academy of Sciences USA*, 103:552-557.

Kivipelto, M., E. Helkala, M. Laasko, T. Hanninen, M. Hallikainen, K. Alhainen, H. Soininen, J. Tuomilehto, A. Nissinen, 2001, Midlife vascular risk factors for Alzheimer's disease in later life: longitudinal, population based study, *British Medical Journal*, 322:1447-1451.

Kolata, G., Years later, no magic bullet against Alzheimer's disease, *New York Times*, 8/28/2010:1.

Krebs, M., K. Domike, and A. Donald, 2009, Protein aggregation: more than just fibrils, *Biochemical Society Transactions*, 37(part 4):682-686.

Landau, L., and E. Lifshitz, 2007, *Statistical Physics, Part I*, Elsevier, New York.

Lei, J., and K. Huang, 2010, Protein folding: A perspective from statistical physics.
arXiv:10025013v1.

Lei, J., S. Browning, S. Mahal, A. Oelschlegel, and C. Weissman, 2010, Darwinian evolution of prions in cell culture, *Science*, 327:869-872.

Lestas, I., G. Vinnicombe, and J. Paulsson, 2010, Fundamental limits on the suppression of molecular fluctuations, *Nature*, 467:174-178.

Levinthal, C., 1968, Are there pathways for protein folding? *Journal de Chimie Physique et de Physicochimie Biologique*, 65:44-45.

Levinthal, C., 1969. In *Mossbauer Spectroscopy*, Debrunner et al. (eds.), University of Illinois Press, Urbana, pp. 22-24.

Levitt, M., and C. Chothia, 1976, Structural patterns in globular proteins, *Nature*, 261:552-557.

Liu, J., J. Faeder, C. Camacho, 2009, Toward a quantitative theory of intrinsically disordered proteins and their function, *Proceedings of the National Academy of Sciences*, 106:19819-19823.

Matusmoto, Y., 2001, *An Introduction to Morse Theory*, Translations of the American Mathematical Society 208, Providence, RI.

Maury, C., 2009, Self-propagating β-sheet polypeptide structures as prebiotic informational molecular entities: the amyloid world, *Origins of Life and Evolution of Biospheres*, 39:141-150.

Meyer, P., 1989, Appendix: a short presentation of stochastic calculus, in *Stochastic Calculus on Manifolds*, Ed. by M. Emer, Springer, New York.

Mirny, L., E. Shakhnovich, 2001, Protein folding theory: from lattice to all-atom models, *Annual Reviews of Biophysics and Biomolecular Structure*, 30:361-396.

Nunney, L., 1999, Lineage selection and the evolution of multistage carcinogenesis, *Proceedings of the London Royal Society B*, 266:493-498.

O,Connor, S., S. Prusiner, K. Dychtwald, 2010, The Age of Alzheimer's, *New York Times*, October 28, A33.

Onuchic, J., and P. Wolynes, 2004, Theory of protein folding, *Current Opinion in Structural Biology*, 14:70-75.

Pappas, G., 1989, *The Magic City*, Cornell University Press, Ithaca, NY.

Petersen, K., 1995, *Cambridge Studies in Advanced Mathematics 2: Ergodic Theory*, Cambridge University Press, New York.

Pettini, M., 2007, *Geometry and Topology in Hamiltonian Dynamics*, Springer, New York.

Plaxco, K., K. Simons, and D. Baker, 1998, Contact order, transition state placement and the refolding rates of single domain proteins, *Journal of Molecular Biology*, 277:985-994.

Protter, P. 1990, *Stochastic Integration and Differential Equations: A new approach*, Springer, New York.

Qiu, C., M. Kivipelto, and E. von Strauss, 2009, Epidemiology of Alzheimer's disease: occurrence, determinants, and strategies toward intervention, *Dialogues in Clinical Neuroscience*, 11:111-128.

Rockafellar, R., 1970, *Convex Analysis*, Princeton University Press, Princeton, NJ.

Royden, H., 1968, *Real Analysis*, Macmillan, New York.

Rudin, W., 1976, *Principles of Mathematical Analysis*, McGraw-Hill, New York.

Sawaya, M., S. Sambashivan, R. Nelson, M. Ivanova et al., 2007, Atomic structures of amyloid corss-β splines reveal varied steric zippers, *Nature*, 447:453-457.

Scheuner, D., and R. Kaufman, 2008, The unfolded protein response: a pathway that links insulin demand with β-cell failure and daibetes, *Endocrine Reviews*, 29:317-333.

Schnabel, J., 2010, Secrets of the shaking palsy, *Nature*, 466:August 26, s2-s5.

Serdyuk, I., 2007, Structured proteins and proteins with intrinsic disorder, *Molecular Biology*, 307:297-313.

Shannon, C., 1961, Two-way communication channels, in Proceedings of the 4th Berkeley Symposium in Mathematical Statistics and Probability, Vol. I, University of California, pp. 611-644.

Sharma, V., V. Kaila, and A. Annila, 2009, Protein folding as an evolutionary process, *Physica A*, 388:851-862.

Sims-Robinson, C., B. Kim, A. Rosko, E. Feldman, 2010, How does diabetes accelerate Alzheimer disease pathology?, *Nature Reviews in Neurology*, doi:10.1038/nrneurol.2010.130.

Singe-Manoux, A., N. Adler, and M. Marmot, 2003, Subjective social status: its determinants and its association with measures of ill-health in the Whitehall II study, *Social Science and Medicine*, 56:1321-1333.

Stern, Y., 2009, Cognitive reserve, *Neuropsychologia*, 47:2010-2028.

Strauss, R., and H. Pollack, 2001, Epidemic increase in childhood overweight, 1986-1998, *Journal of the American Medical Association*, 286:2845-2848.

Tang., M., P. Cross, H. Andrews, D. Jacobs, S. Small, K. Bell, C. Merchant, R. Lantigua, R. Costa, Y. Stern, R. Mayeux, 2001, Incidence of AD in African-Americans, Carribean Hispanics, and Caucasians in northern Manhattan, 2001, *Neurology*, 56:49-56.

Taylor, J., 1996, *Introduction to Measure and Probability*, Springer, New York.

Thayer, J., and R. Lane, 2000, A model of neurological integration in emotion regulation and dysfunction, *Journal of Affective Disorders*, 61:201-216.

Tlusty, T., 2007a, A model for the emergence of the genetic code as a transition in a noisy information channel, *Journal of Theoretical Biology*, 249:331-342.

Tlusty, T., 2007b, A relation between the multiplicity of the second eigenvalue of a graph Laplacian, Courant's nodal line theorem and the substantial dimension of tight polyhedral surfaces, *Electrical Journal of Linear Algebra*, 16:315-324.

Tlusty, T., 2008a, Rate-distortion scenario for the emergence and evolution of noisy molecular codes, *Physical Review Letters*, 100:048101-048104.

Tlusty, T., 2008b, A simple model for the evolution of molecular codes driven by the interplay of accuracy, diversity and cost, *Physical Biology*, 5:016001.

Tlusty, T., 2008c, Casting polymer nets to optimize noisy molecular codes, *Proceedings of the National Academy of Sciences of America*, 105:8238-8243.

Tlusty, T., 2010a, A colorful origin for the genetic code: information theory, statistical mechanics and the emergence of molecular codes, *Physics of Life Reviews*, 2010;doi:10.1016/j.plrev.2010.06.002.

Tlusty, T., 2010b, Reply to comments, *Physics of Life Reviews*, 7:381-384.

Tlusty, T., 2010c. Personal communication.

Tompa, P., P. Csermely, 2004, The role of structural disorder in the function of RNA and protein chaperones, *FASEB Journal*, 18:1169-1175.

Tycko, R., 2006, Molecular structure of amyloid fibrils: insights from solid-state NMR, *Quarterly Reviews of Biophysics*, 39:1-55.

Ullmann, J., 1988, *The Anatomy of Industrial Decline*, Greenwood-Quorum Books, Westport, CT.

Uversky, V., C. Oldfield, A. Dunker, 2008, Intrinsically disordered proteins in human diseases: introducint the D2 concept, *Annual Reviews of Biophysics*, 37:215-246.

Wallace, D., Wallace, R., 1998, *A Plague on Your Houses*, Verso, New York.

Wallace, D., and R. Wallace, 2000, Life and death in Upper Manhattan and the Bronx: toward an evolutionary perspective on catastrophic social change, *Environment and Planning A*, 32:1245-1266.

Wallace, D., R. Wallace, and V. Rauh, 2003, Community stress, demoralization, and body mass index: evidence for social signal transduction, *Social Science and Medicine*, 56:2467-2478.

Wallace, R., D. Wallace, J. Ullmann, and H. Andrews, 1999, Deindustrialization, inner-city decay, and the diffusion of AIDS in the United States, *Environment and Planning A*, 31:113-139.

Wallace, R., and M. Fullilove, 2008, *Collective Consciousness and its Discontents*, Springer, New York.

Wallace, R., and R.G. Wallace, 2008, On the spectrum of prebiotic chemical systems: an information-theoretic treatment of Eigen's paradox, *Origins of Life and Evolution of Biospheres*, 38:419-455.

Wallace, R., and D. Wallace, 2005, Structured psychosocial stress and the US obesity epidemic, *Journal of Biological Systems*, 13:363-384.

Wallace, R., and D. Wallace, 2008, Punctuated equilibrium in statistical models of generalized coevolutionary resilience: how sudden ecosystem transitions can entrain both phenotype expression and Darwinian selection, *Transactions on Computational Systems Biology IX*, LNBI 5121:23-85.

Wallace, R., and D. Wallace, 2009, Code, context, and epigenetic catalysis in gene expression, *Transactions on Computational Systems Biology XI*, LNBI 5750, 283-334.

Wallace, R., and D. Wallace, 2010, Cultural epigenetics: on the heritability of complex diseases, in press, *Transactions on Computational Systems Biology*.

Wallace, R., 2005, *Consciousness: A mathematical treatment of the global neuronal workspace model*, Springer, New York.

Wallace, R., 2007, Culture and inattentional blindness, *Journal of Theoretical Biology*, 245:378-390.

Wallace, R., 2009, Metabolic constraints on the eukaryotic transition, *Origins of Life and Evolution of Biospheres*, 39:165-176.

Wallace, R., 2010a, A rate distortion approach to protein symmetry, *BioSystems*, 101:97-108.

Wallace, R., 2010b, A scientific open season, *Physics of Life Reviews*, 7:377-378.

Wallace, R., 2010c, Extending the modern synthesis, *Comptes Rendus Biologies* 333:701-709.

Wang, L., S. Maji, M. Sawaya, D. Eisenberg, and R. Reik, 2008, Bacterial inclusion bodies contain amyloid-like structures, *PLOSBiology*, 6:e195.

Wallach, J., and M. Rey, 2009, A socioeconomic analysis of obesity and diabetes in New York City, *Public Health Research, Practice, and Policy*, Centers for Disease Control and Prevention, http://www.cdc.gov/pcd/issues/2009/jul/08$_0$215.*htm*.

Weinstein, A., 1996, Groupoids: unifying internal and external symmetry, *Notices of the American Mathematical Association*, 43:744-752.

Wilkinson, R., 1996, *Unhealthy Societies: The Afflictions of Inequality*, Routledge, London and New York.

Wilson, R., Barnes, L., Bennett, D., Li, Y., Bienias, J., Mendes de Leon, C., Evans, D., 2005, Proneness to psychological distress and risk of Alzheimer disease in a biracial community, *Neurology*, 64:380-382.

Wolynes, P., 1996, Symmetry and the energy landscapes of biomolecules, *Proceedings of the National Academcy of Sciences*, 93:14249-14255.

Yang, W., and M. Gruebele, 2004, Folding λ-repressor at its speed limit, *Biophysical Journal*, 87:596-608.

Zhang, Q., Y. Wang, and E. Huang, Changes in racial/ethnic disparities in the prevalence of type 2 diabetes by obesity level among US adults, *Ethnicity and Health*, 14:439-457.

Index